D0445094

Solar Homes
and Sun
Heating

Books by George Daniels

The Getting Game

Home Guide to Plumbing, Heating and Air Conditioning

How to Be Your Own Home Electrician

The Unhandy Handyman's Book

How to Use Hand and Power Tools

The Practical Home Handyman

The Easy Way to Make and Remodel Your Own Furniture

How to Build or Remodel Your House

Make Your Own Monstrosities—With Tooth and Nail

SOLAR
HOMES
and SUN
HEATING

George Daniels

HARPER & ROW, PUBLISHERS
New York   Hagerstown
San Francisco   London

Designed by C. Linda Dingler

Library of Congress Cataloging in Publication Data

Daniels, George Emery, 1914–
  Solar homes and sun heating.

  Includes index.
  1. Solar heating. 2. Solar houses. I. Title
TH7413.D36   697'.78   74–15818
ISBN 0–06–010937–8

77 78 79 10 9 8 7 6 5

Contents

Solar Homes and Sun Heating

1 The Windows That Heated a House

Although the basic idea of solar heating can be traced back to ancient Greece, wide public awareness of its possibilities probably stems from a cold day in the early spring of 1933, during preparations for the Chicago World's Fair. The interior of the exhibit termed "House of Tomorrow" was being completed by a crew working in shirt-sleeves as architects George Fred Keck and William Keck looked on. Although there was no heating system in operation, the house, with its south-facing picture windows, was much too warm for heavy work clothes. The Kecks, long interested in the potentialities of solar energy for supplementary heating, had of course seen other un-planned examples of it. But the House of Tomorrow happened to be the one that triggered their quest for precise data on the heating power of sunlight. Such data would enable them to calculate in advance just how much fuel could be saved by using any given area of south-facing windows to capture the heat of the sun. The data would make it possible to design a house that would have not only the advantage of bright, sunny rooms in winter, but a predictable amount of free heat.

At the time, however, useful solar-energy information was not to be found in the usual sources, such as the established heating and ventilating guides. So the Kecks looked to the U.S. Weather Bureau, which in turn led them to the University of Chicago's weather station. There, a program of tests was under way that recorded solar radiation in gram calories per square centimeter on an hourly basis in the Chicago area.

The Kecks converted the university's measurements from gram

calories to British thermal units (Btu), commonly used in home heating and insulating computation, and were able to calculate that a properly designed house with a south wall made predominantly of glass could bring a fuel saving of 15 to 20 percent—an estimate soon corroborated in houses designed by the Keck firm and built in the Chicago suburbs. The houses performed so well, in fact, that their number grew in one locale to what was often referred to as a solar-house colony. The very term "solar house" appears to have been coined at that time and was used in print, probably for the first time, by the Chicago *Tribune*'s real estate editor, Al Chase.

Ironically, many architects had previously hesitated to use picture windows in homes built in northern climates because of expected heat losses through the large areas of glass. Yet the Kecks' houses, with their walls of windows facing south, not only overcame the losses but frequently took over the entire heating job during the daylight hours, even in subzero weather.

Not surprisingly, interest in the solar house grew rapidly. And industrial advances with a direct bearing on solar-house construction were reaching the production stage. One of these was Thermopane, a double-pane insulating glass with a sealed air space between the panes. How much could it add to fuel economy? The broad facts of solar heating had been known for many years, but the time was ripe for a practical evaluation based on actual year-round performance of a solar house. In 1941, under the sponsorship of the Illinois Institute of Technology and with the financial support of the Libbey-Owens-Ford Glass Company, makers of Thermopane, a year-long solar-house test began. The final results made worldwide headlines. The house selected for the study, designed by George Fred Keck, was built in Chicago's Flossmoor area.

The test house was a one-story frame type with single-slope low-pitched roofs. These extended into wide eaves on the high south side to provide the overhang necessary to shade the windows completely in summer, when the sun is high in the sky, and to allow the low-positioned winter sun to penetrate as far as possible into the major rooms, as shown in the sun-position diagram. (The design not only adapts well to solar heating, but is economical to build and well

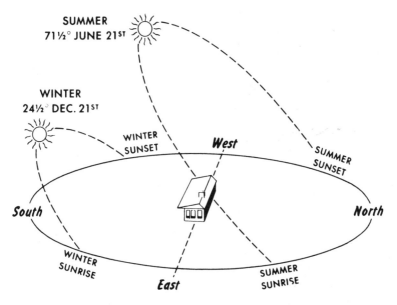

The sun's path through the sky, as diagramed above, can automatically turn on your solar heating in winter and turn it off in summer. The next two photographs show how this actually works in one type of solar heating.

suited to do-it-yourself construction. The same basic concept is used in other houses shown in later chapters. The house in which this book was written is one of them.)

All windows in the test house were glazed with the newly introduced double-pane insulating glass. The living room, dining room, kitchen, bedroom, and study had southern solar window exposures with a total area equal to 50 percent of the floor area. Only the bathroom, dressing room, utility room, and storage areas were on the north. (If you plan a solar house, arrange your layout so that the most-used rooms are sun-heated directly. If your backup heating system is a warm-air type, the blower can carry sun-heated air to north rooms when it's needed, minimizing actual furnace operation.)

The air space within the test-house walls and roofs was insulated with a stitched fiber-filled blanket faced with aluminum foil. (Today, a comparable aluminum-foil-faced fiberglass might be used.) The floor, a 4-inch concrete slab, was poured on a 10-inch bank-run gravel base. Wrought-iron pipe ¾ inch in diameter, spaced 9 inches center to center in coils, was embedded in the slab to carry the hot water of the gas-fired radiant-heating system that heated the house when the sun could not. One such heating panel was located under each of the major rooms with the exception of the storage room. A ⅙-horsepower electric motor drove the pump that circulated water from the boiler through the radiant system. Thermostatic controls governed the heating system in the conventional manner, using an inside-outside temperature control to enhance performance.

Every effort was made to assure the accuracy of the study made of the performance of the test house under everyday conditions. Its owners, a young married couple, lived in it during the entire year-long period of the study, following a normal routine, entertaining their friends, adjusting their heating arrangements as they saw fit, and taking their own notes on its performance. (They became solar-heating enthusiasts.)

Recording instruments kept track of the inside temperatures a few inches from both floor and ceiling, also inside and outside on the north and south sides of the house, while contact thermometers provided a record of wall temperatures.

The conventional heating system was automatically monitored with equal precision. This included the on-off times of boiler operation, the total cubic footage of the gas used, even the temperature of the water entering and returning from the radiant-heating coils. Outside, periodic readings of solar energy were made by special instruments.

All variables could not be calculated, as in instances when the overlapping effect of the conventional heating and the solar heating overheated the house enough to require opening windows. The outdoor element of the automatic control that responded to solar heat had to be modified, as the cold outside air, under certain conditions, proved to have more effect on it than the sunlight, with the

Inside the Chicago-area solar house described in Chapter 1. This photo (taken in midwinter) shows how the sun, low in the sky, shines through the south window wall to keep the house comfortably warm even in subzero weather, with the conventional heating system shut off.

The same room on a sunny day in midsummer. No direct sunlight enters the south window wall because the rays of the sun (high in the sky in summer) are blocked by the roof overhang. Details for planning and building to achieve this automatic effect are found in Chapters 3 and 6.

result that the boiler sometimes continued to operate when the solar energy could easily have kept the house temperature at the required comfort level. The modification, suggested by Keck, consisted simply of placing a milk bottle over the control. The bottle served as an enclosing window for the control, more closely approximating the effect of the solar heat inside the house.

The calculated seasonal heat loss for the test heating season was .6 therms per square foot of window area. (A therm is 100,000 Btu. Details of these heat-measuring units are found in Chapter 2.) The solar-heat input through the window area during the test season based on a study of Weather Bureau data was 1.7 therms per square foot, for a gain of 1.1 therms, showing an overall gain of about 18 percent. Comparing actual fuel consumption to calculated heat losses plainly indicated that the house received at least 12.5 million more Btu than was contained in the total amount of fuel burned.

But to newsmen the highlight of the final report was not so much in the figures as in the summary of performance on a January day in 1942, a day that ranked among the coldest the Chicago area had experienced. The temperature outside hovered between 17 and 5 degrees *below* zero, never higher. Yet at an inside temperature of 72 degrees, controls shut off the gas-fired heating system of the Keck house at 8:30 in the morning and didn't turn it on again until 8:30 at night. Throughout the entire day, although no fuel was used, the temperature inside the house never fell below the pre-set 72 degrees. The sun pouring through that wall of south windows had taken over the job of heating the house, and heating it thoroughly, under frigid weather conditions that Chicagoans would not soon forget. Understandably, the account of that demonstration was reported on radio nationwide, and in the pages of major publications across the country, from *Reader's Digest* to the *New York Times*. The solar home was no longer a mere style or a dream of the future. It was a practical money-saving reality that had proved itself even under the most severe conditions. Sun heating was here to stay. Or so it seemed.

But with the end of World War II still more than three years away, and home building at a virtual standstill, interest in solar heating evaporated. When peace returned after V-J Day in 1945, material shortages were the homebuilder's big problem. Heating fuel, however, was soon cheap and plentiful again, as were the easily installed automatic systems that burned it. The idea of using the sun to heat a house was once again in limbo.

Today, of course, the picture is very different. Fuel is not cheap, and the shortages we have already experienced can happen again. So we can no longer afford the luxury of ignoring the sun as an everlasting source of free heat. No utility company can send you a bill for it, nor can anybody terminate your service. And so far, the only tax on sunlight was an indirect one imposed on the windows of British homes in centuries past, and fortunately repealed long ago. But political intrusions into the field of solar energy bear watching if we are to utilize it freely. As this is being written, "fuel saving" regulations are being proposed in at least one state that would ban the use of glass, even insulating glass, for more than half the area of any house wall. If the proposed regulation should become law, homebuilders would be forbidden to take advantage of the type of heat gain and fuel saving demonstrated so dramatically more than thirty years ago by the Kecks. (Legislators, however, seem to have a penchant for hasty meddling in matters of science, especially when that science is in an early stage of development. In 1910, for example, a bill was considered in Missouri, ostensibly in the interest of safety, that would have banned any airplane from flying higher than 1,000 feet over that state. Even this proposal was more liberal than that of a Louisiana sectarian group that sought to ban aircraft altogether.)

The large-area south window, of course, is just one of many means of utilizing solar energy to heat a house. Some systems, as you'll see in later chapters, can do an efficient heating job even with no south windows at all. Different systems can be used in combination to suit individual situations.

Just how much of the total house heating is taken over by the sun depends on several factors, including your geographical location and your budget. (You can get a rough idea of solar heating's effectiveness in your locality from observations described in the next chapter.) In mild-winter areas you can build a sun-heating system to take

Another type of solar heating system. Rooftop solar-energy collectors on the University of Delaware's Solar ONE prototype home. The house is designed to obtain 80 percent of its heat and electricity from the sun. Further details in Chapter 2.

over the entire heating job at a reasonable cost, though it may be higher priced than a conventional fuel-burning system. Once completed, however, it does something the conventional system can never do. If it's a modern system that "stores" extra heat for use at night and on cloudy days, it can eliminate your fuel bills. More about heat storage later. In areas where temperatures occasionally drop below the mild level, you'll need a conventional backup heating system (like a small oil burner) that can take over when the sun system can't quite keep the house temperature up to comfort level. A number of fringe-area sun-heat systems like this, with conventional backups, have been in operation for years in locations around Washington, D.C., some providing as much as 95 percent of the house heat, often with an annual fuel cost of less than ten dollars at pre-shortage prices. Even if the fuel price increases tenfold, systems like that won't cost much to operate. You can buy plans for one of these (a patented type) and a license to build it. The details are given in Chapter 3.

In cold-winter areas like New England, it's usually wiser at the present time to plan your sun-heating system to handle less than the full heating load. This holds the initial cost down to a figure that can be written off by fuel savings in a reasonable length of time. In operation, a supplementary sun-heating system may do the entire heating job during the milder portions of the heating season, and require backup assistance only during severe cold. Whatever you save on fuel can be used to amortize your sun-heating system. (Typical sun-heating systems handle from 15 to 70 percent of the total house heating job. Later in the book you'll find aids to deciding on the size of sun-heating systems, and the type, along with construction information.) When the installation is paid for, the savings go into your pocket. If possible, plan supplementary sun heating so it can be expanded or modified later. This is especially true of the flat-plate collector types, which are often installed on the roof (sometimes on the lawn). You can build these collectors yourself, as described in Chapter 6, or, for a somewhat higher expenditure, buy them ready-made. When demand increases enough to justify mass production, the price is likely to drop considerably. If it does, and you've planned for expansion of your collector area, it may not cost too much to enlarge your initial sun-heating system to take on all or nearly all of the heating job. Some of the industrial heat-exchanger forms shown in Chapter 6 are already being used as solar-heat collectors, and lend themselves to mass production. As the rest of the system (the assembly that utilizes the heat from the collectors) consists of conventional plumbing and heating materials, it can be made from mass-produced components.

If you're not a do-it-yourselfer or if you can't spare the time to tackle the job of planning and building your own solar home or sun-heating system, you can retain an engineer who specializes in solar homes and sun-heating systems, and have the whole thing

South wall of Chicago-area solar home, as it appears today. Posts support roof over-hang, which blocks sun in summer, admits it in winter. Windows are Thermopane for maximum insulating effect.

designed and built for you. Naturally, you'll pay more, but probably not as much as you might expect. A great many conventionally heated homes are professionally designed and built to order for their owners. And with engineering help, of course, you reduce the chance of problems with solar-heating design.

There are several ways of finding a professional to design a residential sun-heating system. If you plan to buy your heat collectors ready-made, the manufacturer may be able to recommend an engineer or architect in your area who is familiar with sun heating. Otherwise, you can check the yellow pages in your local phone book under Engineers, also under Architects. If some of the engineers' listings mention heating or air conditioning as a specialty, your search may be shortened. You can also write to the American Society of Heating, Refrigerating & Air-Conditioning Engineers (ASHRAE) at

United Engineering Center, 345 East 47th Street, New York, N.Y. 10017, and ask for the address of the organization's chapter nearest to your location. That chapter may be able to direct you to one or more professionals in the solar-heating field who work in your general area. The consultant's fee and other details should, of course, be covered before any work begins.

Don't be misled by the fact that solar homes and sun-heating systems are still relatively few in number. This is by no means due to poor performance or to lack of know-how in their construction, but to simple economics. Their initial cost is still higher than that of their conventional counterparts, for a number of reasons. For one, many of their components are still not produced on a large enough scale to bring the price down. Too, some of the materials and parts required, like the copper sheet and pipe widely used in heat collectors, and the large heat-storage tanks, are high-priced regardless of production methods. And a backup heating system, if required, adds more cost. A final factor is orientation. This calls for careful planning but doesn't necessarily affect cost. The essential point is that of

View of south wall of same house from east end. Look carefully and you can see the angled shadow line on protruding wall. This shows summer sun position.

"tailoring" your solar home or sun-heating system to the individual setting. The heat-gathering elements, whether windows, fluid-circulating heat collectors, or heat-absorbing concrete walls, must be placed for maximum exposure to the sun if they are to have full efficiency. You have to plan your house according to compass direction, not street direction. And you have to plan it so its sun-heating elements won't be obstructed when you build, or later, in landscaping or adding to your home. Also, build your home as far as is feasible from the southern property line to prevent adjacent homes or landscaping from blocking your sunlight at some future time.

On a large building site you may be able to use windows, walls and roof to gather heat from the sun. On a small lot you may be limited to the roof. But with the right system, the roof alone can do the job. Each sunlit square yard of that roof receives an average flow of solar-heat energy about equal to that of an electric hot plate. So a few square yards can provide enough heat for a fair-sized room. A 25 × 40-foot area of sunny roof can deliver a year-round average of better than a million Btu per day—more than most conventional units are ever called upon to produce.

2 How Sun Heating Works —the Basics

The sunlight that shines on the average house in winter carries enough solar energy to heat that house. That's the basic premise of sun heating, and it applies throughout most of the United States and other areas of similar climate and latitude. You can demonstrate it on a small scale merely by placing a dark object in the path of sunlight coming through any south-facing window. Rest your hand on the object a few minutes later, and you find it noticeably warm to the touch. If it's metal, it's likely to be downright hot. (Enclosed in a glass-covered candy box, a piece of black-painted sheet metal can reach 150 degrees or more in minutes on a sunny day.) However small the object may be, it is collecting heat from the sun and radiating it inside the house—actually helping to heat the house. If the window area is large enough, the same general process can do the entire house-heating job during the daylight hours, as in the solar house described in the preceding chapter. But residential sun heating is by no means limited to the south-window approach. Many of today's sun-heating systems require no windows at all (see examples in Chapter 3), although south windows are often used in combination with them. Probably the most popular of these systems is the flat-plate solar-heat collector system, which collects sun heat outside the house and transports it inside by means of a "transfer medium" such as flowing water. This system is used as an example in much of the explanatory matter in this chapter, as it demonstrates many sun-heating principles.

Candy box with black-painted copper sheet in bottom, kitchen oven thermometer resting on it, under propped-up glass cover. Temperature in direct sunlight reached 155 F in minutes. Books hold the assembly at correct angle to sun.

How a flat-plate collector works

In its commonest form, the flat-plate collector is a gridwork of closely spaced pipes bonded to a sheet-metal backing called the plate. Both the plate and the pipes are usually of copper, because it has the highest heat conductivity of the common metals, except for silver, which costs too much for heating-system construction. Another advantage of copper lies in the ready availability of standardized copper plumbing tubes and fittings in the types and sizes suited to solar-heating use. Connections can therefore be made in the same way and with the same tools and solder as in conventional plumbing. The entire assembly is painted flat black, for maximum heat absorption, and mounted (usually on the roof) in an insulated enclosure with a glass top. (Details on building flat-plate collectors and buying ready-made ones are given later in the book.) Because sun heat is con-ducted from the plate to the pipes, the entire area of the collector contributes its heat to the water flowing through the pipes. From there on, the system works much the same as a conventional hot-water heating system, with the heat collector taking the place of a fuel-fired boiler. The principle of heat behavior is the same. Heat always travels from an area of higher temperature to an area of lower temperature. Hence, in the heat collector, sun heat from the plate and pipes travels to the cooler water flowing through the pipes, heating the water. Once inside the house, the hot water gives off its heat to the cooler surroundings, heating the house. So, broadly speaking, the major difference between conventional fuel-burning boiler systems and a sun-heat collector system is where the heat comes from.

How much sun heat can you collect?

Although the roof can usually gather enough solar energy to heat the house, covering the entire roof with sun-heat collectors is likely to cost too much to be practical. So you need to estimate the mini-mum collector area that will do the heating job, or (more likely) a reasonable part of it. To estimate that, you must first know just how much sun heat *per square foot* actually reaches the earth each day during the heating season, and that figure varies with geographical location and local conditions. Naturally, you can expect more sun heat in Florida than in Alaska. And because of air pollution, there may be as much as 15 or 20 percent less in cities than in suburbs. So the important figure is the one that applies to your general locality. This is called the *insolation* (spelled with an *o*) for that locality. Don't confuse it with insulation. Measurements of insolation made by mea-suring stations in different parts of the United States and around the world have been recorded on insolation maps covering the various geographic regions, including yours. Because of the limited number of measuring stations, the maps can't show strictly localized differences, but they can serve as broad area guides adequate for estimating purposes. Maps showing typical monthly insolation throughout the United States and other parts of the world—*World Distribution of Solar Radiation,* by Smith, Duffie and Löf—are available at a moderate charge from the University of Wisconsin's Engineering Experiment Station, 1500 Johnson Drive, Madison, Wisconsin 53706. The National Weather Service also issues insolation maps carrying measurements applying to the United States.

The sunlight-measuring units

The measuring unit used on the insolation maps is the Langley, named for an American scientist whose pioneering work in radiant-heat measurement included the invention of the bolometer, an instrument capable of detecting differences as minute as one ten-thousandth of a degree Fahrenheit. Technically, the Langley is 1 gram calorie per square centimeter, a gram calorie being the amount of heat required to raise 1 gram of water 1 degree centigrade. By tying the amount of heat to the specific area of 1 centimeter, the Langley creates a simple unit for measuring solar energy. For standardization in mapping work, the measurements are taken on a *horizontal* surface. The figures given on the map are not based on a single measurement, but on the *mean daily insolation* for the month designated on the map. An important point: As the measurements are taken on a horizontal surface (on which the sun shines obliquely), you can expect a somewhat higher actual input on a solar-heat collector that is properly inclined to the sun, as explained later.

For house heating, a different measuring unit

When insolation measurements are to be used in house-heating calculations, the Langleys are usually converted to British thermal units

(Btu), the units commonly used in heating work, solar or conventional. The Btu is the amount of heat required to raise the temperature of 1 pound of water 1 degree Fahrenheit. (This amounts to 252
calories—about the amount of heat energy in a couple of good
highballs.) As this unit indicates only the amount of heat (not the
area), it must be expressed in Btu *per square foot* when converted
from Langleys. To convert, you need only multiply the number of
Langleys by 3.69. Thus, 100 Langleys per day (a low reading) equals
369 Btu per square foot per day. To avoid figuring, you can use the
conversion table to convert the map readings.

Once you know the number of Btu per square foot per day that
are available in your locality, simple arithmetic tells you how many
square feet of heat-collecting area you need to provide the total

Conversion from Langleys per day to Btu per square foot per day

Langleys/day	Btu/(sq ft day)
100	369
150	554
200	738
250	922
300	1,110
350	1,290
400	1,475
450	1,660
500	1,845
550	2,130
600	2,220
650	2,400
700	2,580
750	2,760
800	2,950
850	3,140

Use this conversion table for rough figuring where increments of 50 Langleys are
applicable.

necessary for heating your house. Certain minus factors must be figured in, as described shortly. You can't increase the heat output of the sun, but you can increase the amount of sun heat you collect, merely by increasing the area over which you collect it. In sun-heating work, knowing your local insolation is like knowing the number of Btu in a gallon of fuel oil for conventional oil heating: from that it's easy to calculate how many gallons of oil you need to produce any total amount of heat, allowing for normal losses, as described later.

Example of map use

Use the February solar-radiation map to familiarize yourself with insolation figuring. If you look at the New England coastal area, you will find an insolation of 250 Langleys indicated on the line that includes the general region where the solar home shown on page 137 is situated. Converted to Btu per square foot, this comes to slightly more than 922 per day (250 × 3.69). For a sample calculation, multiply 922 by the number of square feet available for sun-heat collecting on the house involved in your plans (or the house you live in). This area might include a south-sloping roof or even a sunny lawn area close to the house, from which insulated pipes could be run. A 25 × 40-foot area (1,000 square feet) would provide a gross total of 922,000 Btu per day with this amount of insolation. If you figure the efficiency of your collectors at 60 percent (typical efficiencies are charted in Chapter 6), you have 553,200 Btu per day remaining. This adds up to more than 149,300,000 for a 270-day heating season—the heat equivalent of more than 1,000 gallons of oil. This is a gross figure, of course, but it gives you an idea of the sun's heating potential.

For a closer estimate, logically, you would work from the insolation figures for each month of the heating season, some of which would be higher, some lower than that for February, but likely to add

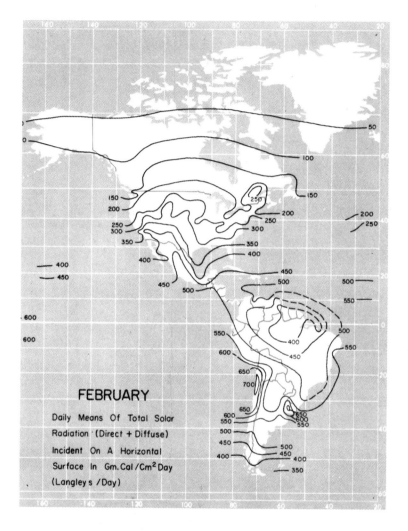

North and South American portion of solar radiation map for February. Numbers in black are readings in Langleys.

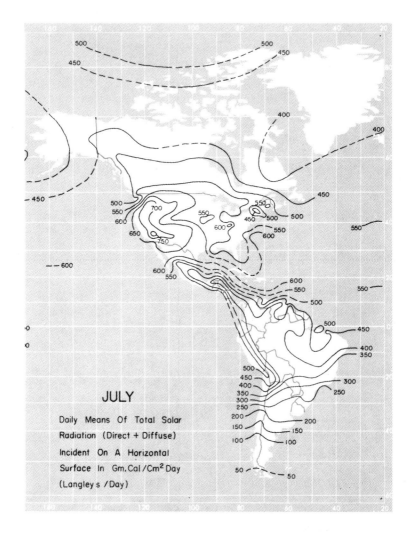

JULY

Daily Means Of Total Solar
Radiation (Direct + Diffuse)
Incident On A Horizontal
Surface In Gm.Cal/Cm² Day
(Langleys /Day)

July readings from Report No. 21 for same area shown in February map. Note that
northern readings are generally higher, southern readings lower, because of seasonal
difference above and below equator.

up favorably. At the start of the heating season in September, for example, the area that showed an insolation figure of 250 in February would be nearer 400. And at the end of the season, in May, it would be close to 500. In the sample region, only January shows a lower reading, around 200. Whatever the reading, approximately 90 percent of the energy is received during the middle two-thirds of the day. Operating the heat collectors at other times produces little or no *usable* heat. This is one of the minus factors mentioned earlier. So you lose a little, but as the original insolation measurements were taken on a horizontal surface, your collector, if properly inclined to the sun, will usually gather somewhat more *total* heat than the map readings indicate. Thus you have a positive factor tending to counterbalance a negative one.

The correct inclination of heat collectors

Generally speaking, a fixed-position heat collector will gather the most heat for its surface area if it is inclined at right angles to the sun's

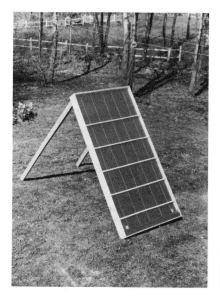

Home-built fluid-type flat-plate solar-heat collector being tested on ground. Details and other illustrations are given in Chapter 6.

rays at noon. As the position of the sun varies from high in the sky in summer to low in the sky in winter, however, you must plan the angle for the season during which the collector will be used most. And the method is extremely simple.

If you'll use the heat collector all year round, as for heating in winter and for swimming-pool heating in summer, plus year-round domestic water heating, you incline it at an angle *equal to your local latitude.* At a latitude of 42 degrees north, as in the Connecticut area, your collector should be inclined 42 degrees from the horizontal.

If you are chiefly interested in sun heating in *winter only,* your collector should be inclined more steeply to be perpendicular to the rays of the winter sun, which is much lower in the sky. So you *add* 10 or 15 degrees to the latitude to determine the correct inclination. If you add 15 degrees (usually preferred) to the 42-degree latitude mentioned above, your collector will be inclined at 57 degrees to the horizontal.

If you intend to use the collector mainly in summer, as for pool heating to extend the swimming season, you *subtract* 15 degrees from the latitude. At 42 degrees latitude, this would give your collector an inclination of 27 degrees. This angle, much closer to horizontal, aims your collector higher toward the sky to match the higher summer sun.

Naturally, the sun's rays are not perpendicular to the collector throughout the day, regardless of the angle setting. Relatively, the sun starts its movement from the eastern horizon, and ''rises'' to maximum height at noon, then moves gradually downward until it ''sets'' in the west. So the actual angle at which its rays reach the collector varies constantly throughout the day, reducing the collection rate as the sun angle moves farther from perpendicular to the collector. This is another minus factor. The inclination angles mentioned above, however, are the most efficient compromise for fixed-position collectors. It's possible, of course, to increase the solar-energy input by building a movable collector that remains perpendicular to the sun all day with an automatic tracking mechanism like those used on large telescopes. But you can gain the same amount of solar heat

much more simply and economically merely by increasing the size
of the collector.

The glass cover

Most flat-plate collectors are covered with a single layer of glass. But
in severe-winter areas, or where peak output is required, two or three
layers, with intervening air space, are sometimes used for their increased insulating and heat-trapping effect. Some energy is lost at
each glass-to-air surface, however, due to reflection and absorption.
A single layer of glass is likely to shut out about 10 percent if it is
nearly perpendicular to the sun's rays. If the light strikes it at shallow,
glancing angles, the reflection is increased and the loss is greater. So
there's a limit to the possible gain, and extra layers are used only
when conditions require them.

The glass may be single-weight window glass unless the possibility
of falling ice or other hazards calls for the added strength of double-
weight window glass. Double-pane insulating glass has been used in
some collectors, as in the Copper Development Association's
Decade 80 solar house in Tucson, Arizona. This type of glass, with
53 percent more insulating ability than single-pane glass, doesn't
"steam up" or collect frost as regular glass does. In conventional
window use with an outside temperature of 30 degrees *below* zero,
the inside surface (with normal room temperatures) is about 37 degrees *above* zero, compared to only about 3 degrees above for
ordinary ⅛-inch window glass, an obvious advantage in heat-collector use. You can figure its insulating value about equal to 8 inches
of concrete. So although it costs more, it's a good choice for heat
collectors in severe-winter areas. (If you want to protect your glass
from breakage by hail or other causes, you can mount ½-inch galvanized wire mesh on a framework a few inches above the glass. To
make up for the reduction in the light reaching the collector, increase
the collector area about 15 percent.) An important point: if you build
a heat collector solely for summer use in swimming-pool heating,

you can omit the glass cover completely, unless the location is a breezy one. You'll cut costs and eliminate all reflection losses.

Heat-wave trapping

The glass cover does much more than just protect the innards of the collector from the weather and keep the heated air from escaping. Because of its highly selective action on sunlight, it allows about 90 percent of the short-wave solar radiation (including the sunlight you see) to pass through and heat objects or surfaces inside the house or heat collector. But it almost completely blocks the resulting long-wave (heat) radiation, preventing its escape. In short, glass, even ordinary window glass, serves as a heat-wave trap. (The commonly available sheet plastics, unfortunately, can't match this long-wave-radiation-trapping ability.)

The flat black coating

The ideal coating for a heat collector would be a "selective absorber" type that would absorb practically all the sun heat reaching it, while radiating little or none of the absorbed energy back to the surrounding atmosphere. In addition, it should be inexpensive and highly durable. Up to now, however, a coating hasn't turned up with all these qualifications. The most selective usually require intricate chemical processes more appropriate to industrial use than to do-it-yourself application. But several paints and inexpensive coatings have proved highly effective. For glassless collectors, one is the black antiglare paint used ahead of the windshield on aluminum aircraft to eliminate glare, and available through aircraft suppliers. Another is Sears' Driveway Coating and Sealer, an inexpensive and easily applied material with good durability. Accelerated one-year tests by its producer, Chevron Asphalt Company, showed practically no loss of energy-absorbing qualities. The important point in using either of these materials: follow the manufacturer's instructions in surface preparation, priming and application. Thorough cleaning of the surface is important. Two coats of the Driveway Coating and Sealer are required, preferably applied early enough to eliminate the chance that dew will form on the coated surface before it has dried for a full day. For glass-covered collectors, Black Velvet paint, manufactured by the 3-M Company, is a good choice.

How much heat will your solar house need?

If you're planning to install a sun-heating system in an existing house, your best guide to heating requirements is last year's fuel bills. If they tell you not only what you paid but how much fuel you received, all you need is simple arithmetic and the heat content of the fuel to find out how many Btu the house requires for the heating season. The approximate heat content of the common fuels is as follows:

Fuel	Btu per unit
No. 2 fuel oil	140,000 Btu per gallon
Natural gas	1,000 Btu per cubic foot
Electricity	3,415 Btu per kilowatt hour

Except for electricity, the Btu content of the fuel units may vary slightly. If you want an exact figure, you can check with your fuel supplier.

For an example of the figuring, let's say your bills show you burned 1,500 gallons of oil during last year's heating season. If you multiply the Btu content per gallon (140,000) by the number of gallons, you get 210 million Btu, which is the amount of heat "input" your house required during the heating season. When you're dealing with figures this large, you may prefer to use the therm as a measuring unit. The therm is a basic heat unit equal to 100,000 Btu. So 210 million Btu is equal to 2,100 therms. If you figure a gallon of oil contains 1.4 therms, you need merely multiply the 1,500 gallons of oil by 1.4 to get 2,100 therms. This simplifies your arithmetic. The same method can be used with other fuels. You'd need 100 cubic feet of gas, for instance, to equal 1 therm (100 × 1,000 = 100,000). So, if you burn gas, you divide your total cubic footage by 100 to determine the number of therms.

The importance of input and output

Although the amount of fuel you burned is your best guide to the amount of heat your house required, you did *not* get all the heat that was in the fuel. Some of it goes up the chimney and there are other minor losses. So oil- and gas-burning furnaces and boilers are tested and rated according to their input and output. The input is the *theoretical* amount of heat in the fuel consumed per hour. The output is the amount of heat *actually produced* per hour. For warm-air furnaces, the term "at the bonnet" is often used in place of the word "output." Its meaning is the same—the actual amount of heat leaving the furnace, before it enters the ducts leading to the rooms. Typical hourly ratings of a small residential furnace are 105,000 Btu input, 84,000 Btu output. Corresponding ratings for a larger size are 140,000 Btu input, 112,000 Btu output. In both examples the efficiency turns out to be 80 percent. Keep in mind that these heating units run intermittently, switching on only when the thermostat calls for heat, and switching off when they have brought the temperature up to the thermostat setting. Running steadily, they'd soon make the house much too hot for comfort. And a residential oil burner with an hourly input of 140,000 Btu might easily consume more than 6,000 gallons during the heating season—enough for a fair-sized commercial building. So figure your home's heating requirements from the amount of fuel actually burned, not on the amount the heating unit *could* burn running steadily.

Electric heat is usually figured at 100 percent efficiency because there is no chimney, and all the heat remains in the house. So you can figure your heat output without a percentage deducted. (You'll have to estimate the portion of your electric bill that applies to heating, as opposed to lighting and appliances.) If your home was built for electric heating, it should also have close to maximum insulation—a boon to solar heating too.

Whatever the type of heating, it's the output that tells the story. When you plan to introduce sun heating in an existing house, you want to know how close your sun-heating system can come to the output figure, while keeping construction costs within reasonable limits. Whatever you save on fuel goes to pay for the initial cost of the sun-heating system, and after that, it's money in your pocket.

If your sun-heating system is to be installed in a new house for which no previous heating figures are available, you can use the methods described in Chapter 4 to approximate the rate of heat loss you can expect, and the amount of heat you'll need to handle it. (Your wisest choice in any event is the best insulation you can afford. For sun heating, the insulating recommendations for electric heating are advisable. This means 3½ inches of insulation, such as fiberglass, in all outside walls and at least 6 inches in the ceiling or in the attic roof, if the attic is heated.) If, before you start your figuring, you'd like to make a very rough estimate, you can use an old rule of thumb sometimes used for oil heat. This calls for 1 gallon of oil for each square foot of heated living area to carry you through the heating season. It's based on an average house in a fairly cold-winter area, but the results can vary widely with location and insulation. In the early pre-planning stages, however, a guided guess is sometimes better than nothing. You complete the figuring, of course, by converting the gallonage to Btu, and subtracting 20 percent to get the output.

How sun heat is stored

As the flat-plate sun-heating system can gather its heat only during the daylight hours, it must store some of it for use at night and during cloudy weather. (The collector can gather considerable heat in hazy weather, and is not dependent on a completely clear sky.) So the modern system includes a large basement storage tank, typically of about 1,500 gallons capacity, in which heat is stored in the form of hot water. In the simplest circulation plan, water (cold at the start of the heating season) is pumped electrically from the bottom of the pre-filled tank through the heat collector and back to an inlet fitting higher on the tank, so the returning hot water mixes with the cold water in the tank. If the collector is of adequate size, the water in the upper portion of the tank reaches house-heating temperature in a matter of hours in clear weather. Hot water is then drawn off by a separate pump from a fitting near the top of the tank (where it's hottest) and circulated through the room-heat-distribution system, then back to the tank. A thermostat shuts off the heat-distribution pump when the room temperature reaches the pre-set level, but the heat-collecting-system pump continues to operate as long as the temperature of the heat collector is higher than that of the tank water. This is done automatically by a differential thermostat with a sensing element in the collector, another in the tank.

If prolonged sunless weather uses up the stored heat, or if the outside temperature drops too low for the sun-heating system to maintain the pre-set temperature, a small conventional fuel-burning backup heating unit is switched on by a thermostat to do the heating job until the sun system can again take over. There are, of course, numerous variations of this system, but the basic principle is the same. To calculate just how much heat can be stored in 1,500 gallons of water, you need a little arithmetic and some elementary physics, which you can pick up as you go along if you've forgotten your classroom work.

The important point of the physics is the specific heat of water. By definition, the specific heat of any substance is the number of Btu required to raise 1 pound of it 1 degree Fahrenheit. The specific heat of water is 1, which means that 1 Btu is required to raise 1 pound of it 1 degree F.

For the calculation, you need some basic facts about water, starting with the number of cubic feet in 1,500 gallons. As 7.48 gallons equals 1 cubic foot, you simply divide 1,500 by 7.48 and find you have 200.5 cubic feet of water. (A pocket calculator is very handy in this and most other sun-heat figuring.) As a cubic foot of water weighs 62.4 pounds, you multiply 200.5 × 62.4 to get a total weight of 12,511.2, or more than 6¼ tons of water in the tank. Since the specific heat of water is 1, it's apparent that you'll need 12,511.2 Btu to raise the temperature of this amount of water 1 degree. And by raising it just that 1 degree you will have "stored" 12,511.2 Btu in the water. If you raise the temperature 100 degrees, easily possible with an adequate heat collector, you will have stored almost *one and a quarter million* Btu. Not all of that heat is usable, however, because that heat, as noted earlier, always travels from an area of higher temperature to one of lower temperature. But you can use the tank water for house heating until the water is cooled down to the room temperature. After that, because there's no difference between the two temperatures, the heat doesn't travel in either direction, and the water can't give off heat to the rooms. Under average conditions, this doesn't occur for a day or two, often more, as the tank temperature is usually higher than 100 degrees.

Extra heat from a rock pile

You can increase your heat-storage capacity by using a larger tank or by using other heat-storage materials in conjunction with the water in the tank. Rocks are one of the least expensive materials you can use, and they can also serve as a means of transferring the stored heat to the air inside the house, as in the patented Thomason system, described in the next chapter. Although the specific heat of ordinary stone is around 0.20, or one-fifth that of water, the stone weighs nearly twice as much as water. So although each pound stores less heat, you can get a great many pounds into a small space.

Fifty tons (100,000 pounds) are used in the Thomason system. This amounts to about three truckloads of egg-sized rocks. Now for the figuring. As you need ⅕ Btu to raise the temperature of each pound 1 degree, a total weight of 100,000 pounds can store 20,000 Btu for each degree you raise the temperature. Raise the temperature 50 degrees and the stones can store a *million* Btu, a sizable addition to your heat-storage capacity. And, as about a third of the space occupied by the stones consists of open space between them, the Thomason system uses them as radiators by driving air through them with a blower. The forced air picks up heat from the stones and carries it to the rooms of the house. (Although metal might seem like good heat-storage material, it isn't, and it costs a lot more than stones. The specific heat of iron, for example, is only 0.11. So a pound of rock can store almost twice as much heat, and a pound of water almost ten times as much.)

A special radiator to heat the house

To do the same heating job as a fuel-fired system, the sun-heat system might require 3 or 4 times as much overall length of base-board radiating units (or other radiators) because its most efficiently operating water temperature is usually lower than that of boiler-heated water. As such an increase in radiator size is usually impracti-cal and uneconomical, flat-plate sun-heating systems commonly transfer the water's heat to the air in the house by means of a fan-coil unit. The fan is normally a blower of the type used in conventional warm-air heating systems. The coil, which resembles a very large automobile radiator, gets its name from the fact that it is made up of many feet of closely finned tubing, coiled into a compact serpen-tine pattern of S-bends and straight runs. For maximum efficiency, several of these patterns may be mounted in the unit, one behind the other, providing a great amount of radiating surface in a compara-tively small space. Air driven through the coil by the blower picks up heat rapidly from the finned tubes, and continues on through ducts to the rooms of the house. Thus you have a combination hot-water and forced-warm-air system. The ducts through which the warm air flows to the rooms are customarily larger than those of a conventional warm-air system because the average temperature of the air in a sun-heat system is usually lower than that in a fuel-fired warm-air system. The duct sizes used for central air conditioning are, in most cases, adequate for a sun-heating system.

If the original heating system is the hot-water radiant type, as with heating pipes built into a slab floor, the sun-heat system can be connected to it in most cases. This eliminates the need for the fan coil and ducts. As a radiant-heating system operates by emitting its heat over a much larger area (than baseboard heating) but at a lower temperature, the water temperature provided by the sun system is high enough for efficient operation.

Water drainage to prevent freezing

The circulating plan described earlier for a typical flat-plate sun-heating system was, as noted, the simplest type. It can be used unmodified in localities where winter temperatures don't drop to freezing. Where freezing temperatures are likely, you can choose either of two methods to keep your heat collector from freezing at night or in sunless weather. If you use the simple basic system already described, you'll have to provide a fail-safe means of draining the water out of it completely whenever outside temperature is likely to drop to freezing. This is frequently done by providing automatic air-inlet valves at the highest point in the system. When the differential thermostat shuts off the circulation pump because of low temperature in the collector, the water drains from the collector back to the tank, and is replaced by air drawn in through the automatically opened air inlet valves. Without the valves, the water would remain in the collector.

The principle is simple enough to demonstrate with a soda straw. Immerse the straw almost full length in a glass of water, then press your finger tightly over the top of the straw to seal out the air. If you now lift the straw from the glass, you will see that the water remains inside. But lift your finger from the top, and the water drains out. If you use the drain-out system, however, you must design and mount your heat collector so there is no chance of water remaining in any part of it. Even an ounce of water trapped in a fitting on a subfreezing night can break the fitting and require immediate repair. To avoid this trouble, all tubes of the collector must have a downward pitch toward the drain-out end. Tubes that slope with the roof, of course, already have ample pitch. Those that run parallel to the roof ridge, however, must also have a slight pitch toward the drain-out end. For the alternative method you need antifreeze.

Antifreeze and the heat exchanger

If you substitute antifreeze for water, you have no worries about the collector freezing. But the price of 1,500 gallons of antifreeze mix to fill the whole system is much too high to be practical. So you turn to a device that lets you use antifreeze in the heat collector, its pump, and its immediate plumbing, without using it in the heat-storage tank. This device is called a heat exchanger. Essentially it's just a long coil of copper or other conductive pipe mounted *inside* the tank at the bottom, with its incoming and outgoing ends *outside* the tank so there can be no intermixing of the antifreeze in the coil with the water in the tank. In practical form it is usually made up of a number of straight lengths of copper tube connected by U-bend fittings at the ends, except for the inlet and outlet, where the tubes connect to fittings that carry them through the end of the tank. The piping from the heat collector and its pump is connected to these ends of the heat exchanger so that hot antifreeze from the collector is circulated through the exchanger and back to the collector in a continuous cycle. An expansion tank or comparable unit allows for expansion and contraction of the antifreeze with temperature changes.

As the hot antifreeze from the collector flows through the exchanger tubes inside the tank, it gives off its heat to the water almost as fast as if it were actually mixing with the water. The rest of the system and its controls remain the same. The pump that circulates the antifreeze through the heat collector and heat exchanger circuit, however, should be of a type suitable for use with antifreeze. The pump manufacturer can specify the correct model. As ethylene glycol, the usual antifreeze, has a lower specific heat than water, the antifreeze solution will hold less heat than water, but not enough to seriously affect performance. But don't use a solution containing more than 60 percent ethylene glycol. A typical heat exchanger in a 1,500-gallon tank might consist of 15 interconnected 10-foot lengths of ¾-inch copper plumbing tube. For more efficient heat exchange for a given tube length, finned and corrugated tubes may also be used, such as those made by the Wolverine Tube Division of Universal Oil Products Company, P.O. Box 2202, Decatur, Alabama 35601. Check the company or your yellow pages for distributors.

If you want your sun-heat system to provide your domestic hot water, you need another, smaller heat exchanger mounted near the *top* of the tank. Small heat exchangers for this purpose are available ready-made from plumbing and heating suppliers. The inlet and outlet ends are outside the tank, as usual. You connect the inlet end to your water-supply plumbing, and the outlet end to the pipes leading to your hot-water faucets. As the cold water from the supply line flows through the exchanger, it picks up heat from the hot water in the tank, and you have hot water for your faucets. If your house is now heated by a fuel-fired hydronic (forced hot water) heating system, the chances are your domestic hot water comes from this type

Finned copper tube for heat exchanger provides much greater heat exchange for given length. This section is a type made by Wolverine Tube Division, Universal Oil Products Company, P.O. Box 2202, Decatur, Alabama 35601.

of heat exchanger. The transfer of heat from the water in your heating boiler is so efficient that hot water for the faucets can be supplied without the use of a separate hot-water storage tank. Such a "tank-less" hot-water system delivers hot water to your faucets as fast as you can use it under normal conditions. But, of course, it requires that the heating boiler be kept hot all year round, which contributes considerably to your annual fuel bill. In summer, the boiler doesn't heat up your house because the circulating pump that supplies the radiators isn't operating. If you use a similar system of domestic hot-water heating in a sun-heating system's storage tank, however, the heat is free.

You can collect too much heat in summer

In summer, when your house is likely to require the least heat, your heat collectors (in most areas) receive the most. So you have to provide a means of reducing their heat collecting. In some systems this is done by draining part of the heat-collecting assembly. This means that the part to be drained must be connected to the overall system in such a way that it can be isolated by valves in order to allow the remaining collector or collectors to continue functioning, supplied as usual by the collector pump. Drained antifreeze should be saved for reuse. The arrangement varies with the individual system, but presents little difficulty. In any event, a temperature-pressure relief valve is essential in each part of the collector system to prevent damage from pressure buildup. Heat collectors can reach steam-producing temperatures quickly if the system's pump is shut off during hot weather without draining the collector. To prevent loss of antifreeze, the relief valve's blow-off pipe should be connected to a tank that has an opening to the atmosphere.

There are other methods of reducing heat collection in summer. In some, heat-reflecting covers are placed over the unneeded heat collectors. The cooling effect can also be increased by providing louvers at the top and bottom of the collectors to permit air to circulate and carry off any accumulated heat. If the heat collectors are mounted on a vertical wall, the heat reduction can be made automatic by providing an overhang above the portion you want to be inoperative in summer. Plan it as you would a south-window sun-heating system.

When air conditioning units suitable for use with sun heat are generally available, of course, summer sun input can be used for this purpose. As described in Chapter 7, units of this type are already in operation on a test basis. So if you hope to add sun-powered air conditioning in the future, plan for it now.

About the rate of flow

The rate at which the water (or antifreeze) flows through your heat collector has a major effect on its efficiency. If it flows too slowly, the heat-transfer coefficient will be reduced and the collector output will be reduced. Also, if it flows very slowly, the water temperature may rise so quickly that within a short distance from the water inlet it becomes so hot that no more heat can be collected, and the rest of the distance it travels is literally wasted. Because the collector then remains at a high temperature, much heat is lost to the lower-temperature surroundings. In short, although the relatively small amount of water flowing out of the collector is very hot, the collector's efficiency is low. When the flow is more rapid, the heat from the collector is spread through a larger volume of water, so the water temperature is lower. But the total amount of heat in it, measured in Btu, is greater, and the collector's efficiency is higher. In a sense, you are using water to cool the heat collector, and if you cool it well you are taking the sun heat out of it more efficiently and putting it where you want it—in your heat-storage tank and heating system.

From the standpoint of heat transfer (assuming sunny weather), the system's efficiency increases as the temperature of the collector decreases with the augmented carry-off of heat. But for practical reasons there's a limit. You want to collect your sun heat efficiently, but at a high enough temperature to heat your house. And you have to do most of your collecting during the middle two-thirds of the day. You also want to avoid temperatures so high as to lose more Btu than necessary to the surroundings. That's why your sun-heating system needs a valve to control the rate of flow and allow you to find the setting that works best.

Where high water temperatures are required, as in operating absorption-type air conditioners in summer, the recirculation of water that is already at a relatively high temperature from passing through the collector builds up to the needed level in a collector designed for such use. As described shortly, the Decade 80 home reaches the necessary water temperature through the use of long continuous tube circuits. The rate of flow can be adjusted to produce the most efficient performance. The water can then "fire" the special cooling units, as detailed in Chapter 7. As air conditioning requires more heat per unit of collector than is needed for heating, it's fortunate that insolation in most areas is higher in summer than in winter. Even so, however, solar air conditioning usually requires greater collector area than solar heating.

The best flow rate, of course, varies with the individual system and the job it has to do. (It also has to be higher with ethylene glycol antifreeze, which, weight for weight, can't carry as much heat as water.) A few examples can serve as possible starting points when you test a do-it-yourself system. The 1,800-square-foot heat collector of the Decade 80 solar house in Tucson, for example, is being tested at flow rates ranging from .45 to .75 gallons per square foot per hour, using ethylene glycol. Multiply the 1,800 square feet by .75 and you have 1,350 gallons flowing through the collector each hour, or 22.5 gallons per minute. The low flow rate is achieved by connecting the tubes in long circuits. Thus, while the flow per tube is relatively high, the flow per square foot of collector surface becomes very low since the same fluid flows over a large portion of the collector surface. This combination produces water temperatures high enough to "fire" two absorption air conditioners with slightly more than 11 gallons per minute each. The cooling units can function at temperatures as low as 190 degrees F, and reach full output at 225 degrees.

A higher square foot flow rate is used in the Sunworks heating system of the house described in Chapter 6, which also uses ethylene glycol. Intended primarily for winter heating, it includes three banks of heat collectors, totaling 400 square feet in area. The overall flow rate is 12 gallons per minute, which figures to 1.8 gallons per square foot per hour. The tubes in this system, however, are connected in parallel grid form, rather than long circuits. The system is designed to use solar energy for about 50 percent of the total heating load.

Design and special features of a collector also affect the flow rate. The Revere collectors described in Chapter 6, for example, are built with rectangular tubes that have more tube area exposed directly to the sun than is the case with round tubes of the usual size, and operate efficiently at higher flow rates. Figured on the basis of a house with about 2,000 square feet of living area, and a collector area of around 500 square feet to be used for heating in a cold-winter region, the flow rate might be on the order of 2½ to 3 gallons per minute per 2 × 8-foot panel. This figures to around 9 gallons per square foot per hour. However, if a large number of tubes were connected in series, as in the Decade 80 home, the nominal flow quantity per square foot would be reduced considerably.

Although flow rates obviously differ markedly with different systems and conditions, they seldom create problems, as a standard high-velocity circulator pump with a ⅙-horsepower 120-volt motor can cover the full range you're likely to encounter. A control valve in the return line from the tank to the pump lets you adjust the flow rate downward. As to the choice of tube circuiting in the collectors, the parallel-grid arrangement allows a wide temperature range simply by varying the flow rate. (The Sunworks collector panels are connected this way.) If you buy your collectors ready-made, check with the manufacturer on other circuits or combinations that might be useful in meeting your particular requirements.

If you are a do-it-yourselfer, you can use your ingenuity to assemble an experimental collector that can first serve as a test unit, then, possibly after some revision, as part of your sun-heating system. Such a unit is shown in Chapter 6. Built partly from new materials, partly from used ones, the unit was modified in several ways, as described, before being given its tryout in practical use.

Why supply tubes are larger

Friction between water flowing through a tube and the inside surface of the tube increases with the rate of flow and the distance of flow, and it's greater in small-diameter tubes than in large ones. As this friction causes a drop in water pressure, it's important to keep your runs of tube to and from your collector as short and direct as possible, and to use adequate-diameter tube for these supply and return mains. For an average system, the diameter shouldn't be less than ¾ inch, which is used in several of the setups described in this book. For longer than average runs, 1-inch diameter might be required. For large-capacity swimming-pool heaters operated by the pool's filter pump, you might need 1½- or 2-inch diameter mains. To further reduce the friction-pressure drop per unit of length, flow velocities in supply and return lines should not be greater than 4 to 5 feet per second.

The effect is illustrated by the pressure drop in tube sizes commonly used in home water systems. The pressure loss due to friction in 100 feet of ¾-inch Type L water tube at a flow rate of 12 gallons per minute would be a little over 15 psi (pounds per square inch). In 1-inch size, it would be only 4¼ psi. If the flow rate is reduced to 5 gallons per minute, the pressure loss in 100 feet of ¾-inch tube comes down to a mere 5¼ psi, and in 1-inch tube, to less than 1 psi. The smaller-diameter tubes in the collector itself don't create a problem, as the total flow rate is usually divided among them, so the flow rate through each in gallons per minute is only a fraction of the total. And the lower flow rate greatly reduces the friction and pressure loss.

In most of your sun-heating work you're likely to be dealing with ½-inch and ¾-inch diameter tubing, or Revere's rectangular tubing, which has a pressure-drop rate about the same as ¾-inch Type K round tubing. Charts covering the pressure-drop rates of other tubing sizes are available from various manufacturers. If you use them for any purpose, be sure to match them to the *type* of tube you're using. The reason: The actual *outside* diameter is always ⅛ inch larger than the nominal diameter—so 1-inch tube is really 1⅛ inches in outside diameter. But the *inside* diameter differs with the *type* of tube because the thickness of the tube wall differs. Of the water-tube types, Type K has the thickest wall, Type L is of medium thickness, and Type M is the thinnest. In 1-inch tube diameter, the wall thickness in inches for Type K is .35, for Type L .30, and for Type M .25. In the types you're most likely to use, the smaller inside diameter of Type K in nominal ¾-inch size would give you a pressure drop of only a little over 19 psi in 100 feet with a flow rate of 12 gallons per minute. Type L, in the same nominal diameter, but with its slightly larger *inside* diameter, would give you a pressure drop of only a little over 15 psi under the same conditions. The lower figure results from an inside diameter only 4/100 inch larger, a good illustration of the effect of diameter.

As to the directness of your piping, each bend adds some friction, which is equivalent to added tube length. A 90-degree elbow in ¾-inch nominal diameter, for example, is the frictional equivalent of 2 feet of straight tube. A 45-degree elbow in the same diameter, however, equals the friction of only 1 foot of tube. So use mild direction changes in preference to sharp ones where feasible, if by so doing you don't increase the total number.

Concentrating heat collectors

As will be shown in Chapter 3, you can concentrate sun heat by means of lenses or reflectors to attain temperatures of several thousand degrees. Although this is far above flat-plate collector temperatures, you are *not* actually collecting more heat. You are starting with the same number of heat units per square foot as you would have on a flat-plate collector, but you are concentrating them on a smaller area and producing a proportionately higher temperature. (You can set a piece of paper afire merely by bringing sunlight to a pinpoint focus on it with a large magnifying glass.) Because of the high cost of the heat-concentrating equipment, and the fact that it must follow the course of the sun, it is used largely for experimental work at present. It is one method of employing solar energy to produce power by the generation of steam, however, and has been applied to creating temperatures suitable for use in absorption refrigeration systems. Air conditioning units of the absorption type now being tested, however, can operate on heat attainable with flat-plate collectors. More about this in Chapter 7. If you want to experiment with heat-concentrating collectors, you can use either parabolic reflectors or Fresnel lenses to keep costs down. Both are available from Edmund Scientific Co., 150 Edscorp Building, Barrington, N.J. 08007.

Predicting sun-heating system performance

To predict the performance of your sun-heating system with maximum precision, you'd need mathematics beyond the scope of this book, including some calculus and differential equations, plus a background in thermodynamics and other fields. Even then, your results would depend on the accuracy of your insolation figures, which is limited by the relatively small number of measuring stations. You can retain a consulting engineer to do your planning if you want this type of precision. Or you can use the basic information, examples and charts in the book to plan on a less complex basis, and obtain your final performance figures from actual tests. As exact performance data on the various sun-heating system types in different geographical areas is still very limited, and as many unpredictable factors are involved, a certain experimental factor remains. To allow for it as well as possible, you can plan to permit some degree of system modification in case it's needed later. Allowing for increased collector area, for example, would enable you to boost your sun system's heating capacity. Naturally, you have more leeway in mild winter areas. And there are two factors that tend to cushion possible miscalculations. First, whatever heat you get from your sun system is free. And you always have your backup system to take over when it's needed.

Backup systems

The type of backup system you use depends on the individual situation. In an existing house already heated by a conventional fuel-fired heating system, it's usually simplest to use that system as the backup when you add sun heating. If you are building a new house, choice of backup system depends on the nature of the sun-heat system. As you will note in Chapter 3, the Thomason Solaris houses use as backup a fuel-fired warm-air furnace, which feeds the warm air directly to the rooms. Others, like the Barber house in Chapter 6, use a large fuel-fired water heater, and feed the backup hot water to the fan-coil unit to provide warm air for the ducts and reserve heat for the storage tank. The important point: the backup system must have adequate capacity to take over the entire heating job when necessary.

About the cost

Because of price fluctuations for materials and components, the only way to determine overall costs accurately is by getting current prices on the materials you'll need for the system you plan. In general, however, you can figure that a sun-heating system capable of taking over a major part of the house-heating load costs from two to three times as much as an oil-fired hot-water system, and possibly four times as much as an electric heating system. One reason for the higher cost of the sun-heating system lies in the fact that in new construction it must include a conventional backup system. Naturally, the overall price also varies with the location. In mild winter regions, where less heat is needed, costs will be lower because less collecting area will be required. In any area, at the present time, it's wise to explore the possibilities of *supplementary* sun heating. A system designed to take over part, rather than all, of the heating load can be built for a price that can be written off by fuel savings in much less time than would be possible with total sun heating. Most of the sun-heated houses in this book were designed for less than 100 percent sun heating, in order to reduce initial cost. As the use of sun-heating systems increases, it is likely that ready-made components such as heat collectors will be mass produced, lowering costs. A supplementary sun-heating system might then be enlarged to take over a larger portion of the heating load.

If you are planning to build a house, it may be possible at little or no increase in cost to allow for sun heating at a later date. Merely by proper orientation, for example, a south-sloping roof area suitable for heat collectors can be provided. And if the building site is suitable, a large south-window area can furnish considerable supplementary sun heating at relatively low cost, as described in Chapter 1. In wall space allowed at both sides of the window area (before room corners), insulating panels of wood-framed foam can be used to cut nighttime heat losses through the window area, giving even greater sun-heating efficiency. (Plans for insulating panels, which are inexpensive to build, are included in Chapter 6.) In any event, whether you own a house or plan to buy or build, look for sun-heating possibilities. Even if you use the sun only to heat the water for your faucets, possibly with as little as 50 square feet of collector area, you may cut your fuel bills as much as 15 percent.

3 Fourteen Ways to Free Heat

Many different types of solar-heating systems have already been tried successfully. Fourteen of the outstanding ones are illustrated and described in this chapter. Some are designed to meet specific requirements, others are planned for simplicity or for minimum cost. If you intend to install a sun-heating system in a new home or an existing one, the basic information on these successful systems can help you make your selection. Keep in mind, too, that many sun-heated homes utilize more than one type of system in order to increase efficiency.

The south window wall

The potentialities of southern orientation to provide heat were first recorded by the Greek writer Xenophon in the fourth century B.C. His suggestion, in liberal translation: Make the south wall higher than the north, thereby providing a large area to absorb the sun's heat and warm the house, while having a much lower north wall "to keep out the winds."

Although Xenophon's thoughts on solar heating were put down more than two thousand years before window walls, the basic principle is used today, as in the Chicago house described in Chapter 1. (Another interesting variation of the idea is found in the Trombe-Michel solar wall, later in this chapter—an easily built solar-heating design likely to find wide use in future solar homes.)

The principle of the window-wall type of solar home is extremely simple. The sun shines through a large area of the south-facing windows, warming objects and surfaces inside the south rooms of the house. Although the window glass freely admits the short-wave components of the sunlight, including the visible wavelengths, it blocks the escape of the long wavelengths of the heat it produces within the house, as explained in Chapter 2. This heat-trapping quality of glass —often termed the "greenhouse effect"—is a major factor in most solar-heating systems.

Effect of double glass with special dry gas between. Single-glass window at left is much colder on inside surface, collects moisture from air inside house, may also frost up. Double-glass window at right has warmer inside surface at same outside temperature, does not collect inside moisture or frost up. Single glass in an insulated home with 30 percent indoor humidity begins to sweat when outside temperature drops to 28 degrees F. Double glass doesn't sweat under same inside conditions until outside temperature drops to −8 degrees. When outdoor temperature is −10 degrees, inside surface of single glass surface is 15 degrees, that of double glass (with dry gas between) is 41 degrees.

Precise efficiency figures for window-wall sun heating are difficult to compile because of the many variables, which include the nature of the home's interior, the amount of insulation, prevailing winds and their velocity, and other factors to be covered shortly. Architects experienced in this type of solar home, however, feel that, properly designed, it can cut heating costs 15 to 20 percent.

In the normal course of events you have probably experienced the possibilities of window-type sun heating. Returning to a parked car that was left in a sunny spot in chilly weather, you probably found it warmer than the outside air. In summer, most of us have noticed that sun heat can make our parked cars much too warm for comfort, so we park them in the shade when we can. Some of the heat, of course, is being radiated into the car from the hot roof. This portion of the heating effect can be compared to the operation of flat-plate solar-heat collectors. To feel the effectiveness of window-type sun heating, however, you have only to sit on your car's sun-baked seat.

The heat is definitely there. We need only utilize it properly. For this we need enough window area to admit the required amount of sun heat, and of course the window area should face generally south to receive its heat during the major part of the day. The south-window area of the highly successful solar home described in the preceding chapter was equal to approximately 50 percent of the floor area—a proportion that can serve as a guide if you choose this type of sun heating. That house was also well insulated, and glazed with double-pane glass (Thermopane), important features to remember.

The overhang

Through the south-facing window wall, the winter sun, low in the sky, bathes the south rooms in sunlight. But in the summer, the sun, now high in the sky, leaves them in cool shade if they are protected by the shadow of a roof overhang, which provides a completely automatic heat shut-off.

The dimensions of the overhang depend on such features in the design of the south wall as the amount of solid wall that must be provided above the windows for structural purposes. A few seasonal measurements can provide a foolproof guide in any given geographic region to the angle between the edge of the overhang and the window wall to be built. As shown in the diagrams, you can find this angle for winter and for summer by measuring the shadow of any vertical post or wall that casts its shadow to the north.

If you plan to build or remodel so as to use this type of sun heating, you can design your overhang on the basis of these measured angles. By measuring at a number of different times of the year, you can know in advance when the shadow shifting will begin and end each cycle, as it takes place gradually. It can also be calculated on the basis of your latitude and the change in sun position, but actual shadow measurement may reveal local details like shadows cast by trees or high terrain. If you plan to build on the east side of a mountain, for example, you can expect an early sunset at any time of year.

As to the house itself, structural elements of considerable mass, such as the brick and masonry fireplace wall and slab floor of the Chicago house (Chapter 1), provide limited heat storage. If, as in that instance, the slab floor also contains radiant-heating pipes supplied

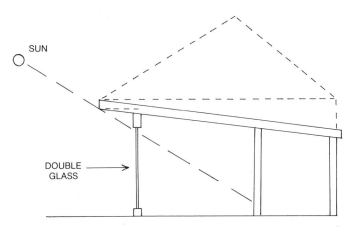

Low winter sun angle of about 24.5 degrees on December 21 lets sunlight enter window wall under roof overhang.

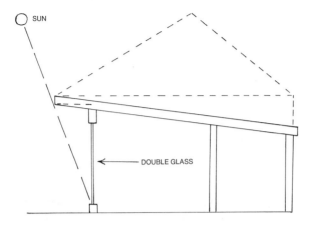

High sun angle of about 71.5 degrees in summer (June 21) results in roof overhang shading window wall, so room is in shade.

by the backup system, it also tends to provide a smooth transition from conventionally supplied heat to sun heat at the beginning of the day. As the house then has a "warm start," many owners of south window walls find that their regular heating plants shut off entirely on sunny days. When flat-plate collectors are used in addition to a window wall to make a combination sun-heating system, true heat storage can be provided by the usual heat-storage tank.

One problem of the window wall lies in heat losses at night and during sunless weather, as even double-pane glass loses heat several times faster than a well-insulated wall. These losses are now being greatly reduced, however, through the use of insulating foam panels (either sliding or folding) that can be moved easily into place inside the window area at night. The panels can be covered with drapery fabric or thin paneling to make them part of the room's décor. The same type of insulating panels in smaller sizes can be used to cut heat losses from conventional windows. Construction details are given in Chapter 6.

In addition to possible heat gain from the sun, the large south-facing window's winter sunshine effect is worthwhile psychologically. And of course it eliminates the need for artificial daytime lighting in most parts of the house, aiding in the reduction of utility bills. A wise approach is to consider the south window wall's advantageous features in combination with another form of sun heating that can provide heat storage.

Roof-reservoir sun heating

This type of sun heating, as in Harold Hay's Skytherm houses, is another uncomplicated system, though it requires considerably greater structural strength than other types. Basically, the system consists of black plastic "water bag" reservoirs resting on a metal roof decking. This is topped by movable panels of insulating foam, fitted with wheels and riding rooftop tracks. On sunny winter days, the insulating panels are moved along their tracks onto the top of an

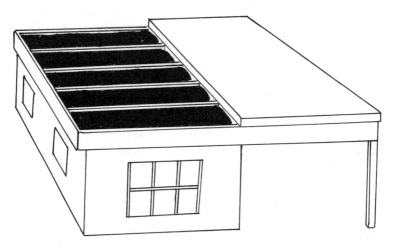

Roof reservoir system developed by Harold Hay uses movable roof insulation and water-filled black plastic bags resting on metal roof. To expose bags to solar heat, insulation is moved away from reservoir area on roller-track system, to area over carport, as diagramed here.

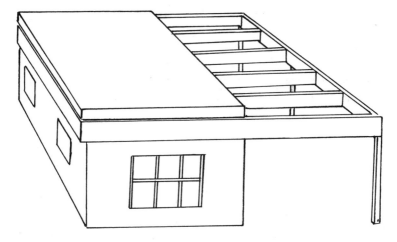

To prevent escape of heat from reservoir at night, insulation is moved back over reservoir, as here. System can be reversed for cooling by exposing reservoir at night to cool air, covering it in daytime.

unheated carport (or similar structure) to expose the reservoirs to the heat of the sun. The heat thus produced in the reservoirs is radiated to the rooms below through their metal ceilings. The excess heat stored in the reservoirs heats the rooms in the same manner during the night, when the foam insulating panels are moved back in place to prevent the escape of the stored heat.

In summer, the process is reversed. The insulation is moved off the reservoirs at night to cool them by radiation to the night sky. It is rolled back at daybreak to prevent the sun from heating the cooled reservoirs, which then absorb heat from the rooms below, providing a natural form of air conditioning. By flooding the plastic-enclosed reservoirs with a thin layer of water (allowed to evaporate during the night), the cooling effect is increased. Tests indicate that the night-time evaporation of around $3/16$ inch of water over the reservoir area can easily cool the reservoirs below the surrounding air temperature and, with the loss of less than 2 gallons of water, produce the effect of a ton of refrigerant.

Performance reports on a test house of this type in Phoenix, Arizona, indicate that in summer the system held inside temperatures between 74 and 77 degrees F though outside temperatures reached 100 degrees. In winter it held inside temperatures between 66 and 73 degrees F when outside temperatures fell to the freezing point. Because of its method of operation, this system is generally limited to snow-free regions with relatively high insolation (solar radiation).

The Trombe-Michel solar wall

This solar-heating system features a south-facing wall of concrete a little over a foot thick, painted black. It is shielded by a double layer of glass spaced with the inner layer about an inch from the outside surface of the wall. When sunlight passing through the glass heats the wall, the air between the wall and the glass is also heated; the heated air expands and rises and, since the top of the air space between wall and glass is sealed, enters the house through openings in the top of the wall. As the warm air flows into the house, it is constantly replaced by air flowing through openings at the bottom of the wall, creating a "gravity" type warm-air heating system requiring no motor and with no moving parts. The bottom openings are located slightly above the bottom of the air-heating space so that at night and on cloudy days, cold air won't flow into the house. Instead it sinks to the deeper bottom of the air-heating space.

In summer the system can provide cooling air circulation. The openings into the house through the top of the concrete wall are closed, and the top of the air space between the wall and the glass is opened to the outside. The heated air then rises and escapes to the outside, drawing air from the house out through the bottom openings in the wall. This air is replaced by air drawn in through openings in the north wall, for a cooling flow.

The system was developed by Professor Felix Trombe, director of the CNRS Solar Energy Laboratory and its related solar furnace in the French Pyrenees. Prototype houses in which the system is now functioning are the work of architect Jacques Michel. Development

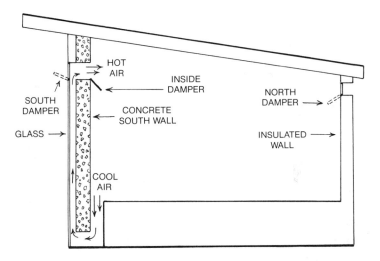

Trombe-Michel solar-heating system uses foot-thick concrete wall facing south and painted black. Sun heats black wall and also air between wall and glass, causing heat-expanded air to rise. In winter, outside south damper is closed, inside south damper opened, so heated air enters house through upper wall openings. North damper is closed. Heated concrete wall stores enough heat to keep house warm through most of night after a sunny day. In summer, outside south damper is opened, inside damper closed, and north damper opened. Heat-expanded air then rises and exits through outside south damper, and is replaced by fresh air drawn in through open north damper. Incoming air is drawn across room, then down through floor register and out, creating cooling breeze through house by natural flow.

is continuing on the prototypes, and may include the use of insulating panels and duct shutters to increase efficiency or meet special conditions. Patentable aspects of the system are protected.

Present indications are that about 1 square yard of the glass-enclosed wall can heat about 10 cubic yards of living space in the house. Though this estimate is based on operation in the prototypes' latitude, it nevertheless illustrates the possibilities of the system. Because of its simplicity, it offers promise for the do-it-yourselfer. Except for the concrete south wall, which serves as a heat-storage element as well as part of the structure, the house may be of conventional, well-insulated frame construction. Small fixed windows might be included in the concrete south wall.

A possible initial do-it-yourself application of the principle: adapting a stone or masonry-walled outbuilding, such as an unattached garage, to simple solar heating.

The Thomason system

The patented solar-heating design developed by Dr. Harry E. Thomason has been tested for fifteen years in homes built by Thomason in the Washington, D.C., area. The system has provided as much as 95 percent of the winter heat for a three-bedroom home. In terms

Thomason house. Roof pitch on left faces south and is the heat collector. Low, nearly flat roof of house extension serves as reflector to increase efficiency of heat-collecting roof. A number of houses using patented Thomason system have been built and are operating successfully.

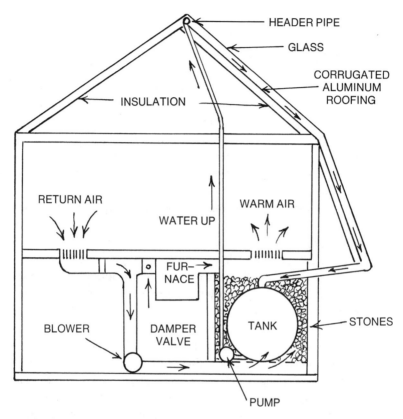

How the Thomason system works. Use of aluminum corrugated roofing to heat water for heating and heat storage reduces costs. This is basic principle. Actual system has various additional features, such as water filter, etc. System can be varied to suit house and climate.

of fuel costs during a sample three-year averaging period ending in 1963, this amounted to bills of $6.30 per winter (pre-shortage prices).

The Thomason system is one of the least expensive solar-heating systems to build, its heat collector consisting of corrugated aluminum roofing (painted flat black) with a glass covering approximately 2 inches above it. At the top of the heat collector, which takes the form of the home's gambrel roof, a pipe within the collector, running parallel to the roof ridge, supplies the water that carries the sun heat into the house. The water (pumped from a 1,600-gallon basement storage tank) is emitted through small holes drilled in the pipe at spacings matching those of the aluminum roofing's corrugations. The water flows down the corrugations of the aluminum roofing, absorbing heat as it goes. At the bottom of the glass-covered roof it flows into a gutter within the collector, and from there down the return pipe into the storage tank from which it came. As the water circulation continues (on sunny days), the entire contents of the tank reach home-heating temperature.

The tank rests on hollow masonry blocks arranged so their hollow cores serve as ducts for air supplied by a blower at one end. The air flows out through spaces left between the blocks, then up through approximately three truckloads of egg-size stones that fill the space between the tank and the masonry-walled "heat bin" in which it is located. The rocks, constantly heated by the tank, provide a large-area heat-exchange medium that warms the air as it flows through them on the way to the room-heating registers.

The water pump is automatically shut off when the temperature of the heat collector drops below that of the water in the storage tank, and the self-draining circulation system empties back into the tank without danger of air locks. So the possibility of nighttime freeze-ups is eliminated. For this reason, no antifreeze is required and no heat-exchanger coil (as is needed for tube-type heat collectors). If the stored water drops below home-heating temperature levels during prolonged cloudy weather, a backup warm-air furnace switches on automatically and supplies heat through the ducts of the solar-heating system.

In some areas the *north* slope of the roof can be used for summer air conditioning by running storage-tank water down it at night through a circulating system similar to that used for winter heating. Without a glass cover, the water is cooled both by radiation to the night sky and by evaporation of a small amount as it flows to the return gutter.

The system can also provide year-round domestic hot water, as well as summertime pool heating. Complete plans, including details of plumbing and controls, for Thomason solar houses with this type of solar-heating system may be purchased from Edmund Scientific Co. (see page 35), which also provides the license required for construction of a house using the patented features.

Fluid-type flat-plate collector systems

This system, widely used in many cold-winter areas, is described in detail in Chapter 2, as it illustrates a number of sun-heating principles. Its general design and operation are repeated here, however, for comparison with the other systems included in this chapter. Its principle of heat collection is somewhat similar to that of the Thomason system, though the water flows through an enclosed gridwork of tubes. In most designs, this gridwork consists of a large number of parallel small-diameter tubes that follow the pitch of the roof. At each end these tubes are connected to a larger-diameter tube called a "header," which runs parallel to the roof ridge. Water pumped from the tank into the lower header (at the eave of the roof) flows upward through the parallel tubes, to the upper header and back to the tank through a return line. (In some designs the direction of flow is reversed, entering the top header first and flowing to the lower one. Efficiency seems about the same either way.) A water tank like that in the Thomason system is used for heat storage. Heat distribution through the house, however, is usually by means of a fan-coil unit, as described in Chapter 2. Blower-driven air, passing through the coil, picks up heat from the sun-heated water flowing inside the tubes of the coil, and carries it through ducts to the rooms of the house.

For cold-winter home heating, as in New England, where the system is to carry from 50 to 75 percent of the heating load, a broad rule-of-thumb approach calls for a heat-collector area equaling 40 to 60 percent of the heated floor area of the house. Relatively smaller collector areas (to 30 percent of floor area) can be used in milder

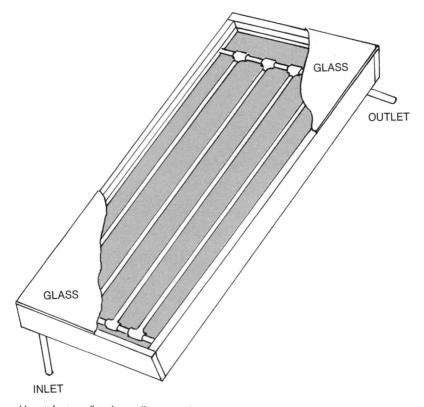

How tube-type flat-plate collector works. Basic idea takes many forms. This one is typical.

climates. All this assumes a well-insulated house with unshaded collector locations.

As shown in the drawing, the tubes are bonded to a metal backing sheet called the plate. As both the plate and the tubes are of a conductive metal (usually copper), heat from the entire collector area is conducted to the water in the tubes, provided that the bond between plate and tubes is also highly conductive. This is very important, as the surface area of the plate exposed to the sun is much greater than that of the tubes. And the conductivity of the bond can vary by as much as 1,000 Btu per hour between a good bond and a poor one. As described in Chapter 6, soldering is one of the common methods of bonding, but heat-conducting epoxy is now used on recently developed flat-sided-tube type collectors. Whatever the bonding method used, the entire assembly is painted flat black to assure maximum heat absorption.

Like the Thomason roof collector, the flat-plate collector is covered with glass when used for winter heating. For swimming-pool heating alone, the glass is often omitted, which reduces the cost of the collector and eliminates any losses due to reflection from glass covers. Complete details of fluid-type flat-plate collector construction and operation are given in Chapters 2 and 6.

Thermosyphon solar water heater

The sun-fired water heater can provide domestic hot water without fuel or power cost (during clear or hazy weather) in any area where sun heating is feasible. A simple electric heating element, made for water heating, can be installed in a lower opening of the tank to act as a backup heat source when required by inclement weather. A thermostatically controlled model should be used.

In areas where winter temperatures do not reach the freezing point, the heater often consists simply of a flat-plate heat collector, typically about 50 square feet in area, and an insulated 40- to 80-gallon tank above it. The collector area, of course, may have to be larger in very-cold-winter regions, or where an unusually large amount of hot water is required on a daily basis. As shown in the drawing, sun-heated water rises by natural convection through a pipe from the top of the collector to an entrance fitting about two-thirds of the height of the storage tank. Cold water from the bottom of the tank is led to the bottom of the collector through another pipe. The bottom of the tank should be set at least a foot (preferably 2 feet) above the top of the collector to prevent reverse flow on chilly nights. Water from the supply line enters the tank near the bottom, and hot water for domestic use leaves it through the top.

In winter-freeze areas, the system is modified to include a heat-

Thermosyphon water heater being tested at Solar Energy and Energy Conversion Laboratory of University of Florida. If tank is placed higher than heat collector, system can function without a pump. Tank can be located below collector if a pump is used to circulate the water.

exchange coil inside the tank. Antifreeze is then used in the heat collector. It circulates by natural convection through the heat-exchange coil inside the tank, transferring its heat to the water in the tank without any intermixing of antifreeze and water. (Additional details on heat-exchange coils are given in Chapter 2.) The tank, of course, must be protected from freezing, along with the piping that carries water to and from the tank. In a house with a steeply pitched roof, the tank can be installed in horizontal position near the roof peak inside the attic. The heat collector may then be mounted lower on the roof outside to provide natural-convection circulation without the problem of nighttime reverse flow. The attic roof should be insulated in this type of installation to protect the tank and its plumbing, all of which should also be insulated. Provision should also be made to keep the attic temperature above freezing.

In the climate common to Israel and Australia in which this type of heater is extensively used, collector areas as small as 16 to 20 square feet and tank capacities of 40 to 60 gallons have been found satisfactory for one-family homes. After a sunny day in winter, these

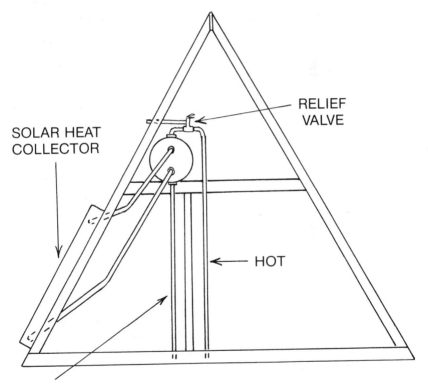

RELIEF VALVE

SOLAR HEAT COLLECTOR

HOT

COLD SUPPLY

How thermosyphon solar water heater works. Water heated in collector expands and rises, leaving top of collector and entering storage tank. As water leaves collector at top, cooler water enters collector at bottom to replace it. Cycle continues until entire contents of tank is hot. As water is used, fresh water enters through supply pipe. Principle is the same regardless of size. Tank should be 2 feet above collector top.

systems can provide a full tank of water at around 120 degrees F. In summer, it's likely to be closer to 165 degrees. Larger collector areas are usually required in the United States, especially in the winter-freeze areas.

Where the tank can't be located above the collector, as in many modern homes, it may be installed wherever space permits, and provided with pump circulation from the collector. The pump system should have a control valve to adjust the rate of flow of the water or antifreeze through the heating circuit. Typical flow rates are around 1 gallon per hour per square foot of heat-collector surface.

Concentrating heat collectors

Where high temperatures are essential, concentrating heat collectors are often used. In one form, they consist of parabolic reflectors that gather sunlight and reflect it to a focal point much smaller than the collector area over which it was gathered. In this way the heat is concentrated from a large area onto a much smaller area, greatly increasing the temperature. The giant solar furnace at Odeillo, France, utilizes this principle to attain temperatures of thousands of degrees. Another form of concentrating collector accomplishes the same end through the use of a lens, commonly of the Fresnel type. In its usual form, this lens is a thin, flat sheet of transparent plastic on which very fine ridges are molded in concentric circles from the center of the sheet to the rim. Each ridge is beveled to refract light in much the same manner as a conventional lens. For light-concentrating purposes the Fresnel lens has the advantage of much lower cost (a very small fraction of the cost of a conventional lens of equal size) and light weight. (At this writing, Fresnel lenses just under 1 foot square are available from at least one scientific supply house for six dollars.) Like conventional lenses, they are made in various focal lengths. In heat-concentrating applications they offer another means of gathering sunlight over a relatively large area and focusing it on a small area to increase the temperature.

Collapsible solar cooker developed at University of Florida. Aluminized paraboloidal umbrella reflects and concentrates solar heat on teakettle or other cooking utensil.

Although heat-concentrating collectors have proved successful in experimental work, in generating steam and operating absorption refrigerating equipment, and in other thermal devices, they have two major limitations. First, in order to deliver a constant flow of heat they must "track" the sun as it moves across the sky. Otherwise the sunlight can't remain focused on the heat receiver. As the tracking mechanism required to do this automatically is both complex and costly, it's not suited to most residential applications. The second drawback to the concentrating heat collector is its inability to deliver heat in hazy weather, as diffuse light can't be focused satisfactorily. (As noted in Chapter 2, the non-concentrating flat-plate collector can

Rigid paraboloidal reflectors. Used for experimental work at University of Florida, these are the type that may be used with sun-tracking mechanisms that keep them aimed at the sun throughout the day. High temperatures are possible with this system, but it doesn't function on hazy days, as diffuse light can't be focused.

deliver a considerable amount of heat in hazy weather, as focusing is not required.) If you'd like to experiment with concentrating solar-heat collectors, you can buy Fresnel lenses and parabolic reflectors in a variety of sizes from a scientific supply house, such as the Edmund Scientific Co. (See page 35.)

Glass-vacuum heat collectors

Still under development by Corning Glass Works, the glass-vacuum type of heat collector consists of two concentric glass tubes with a relatively high vacuum in the sealed space between them. The

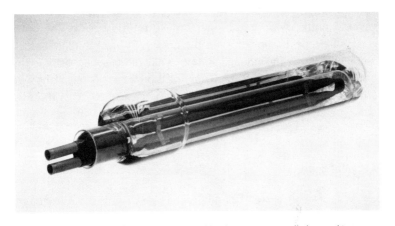

Corning Glass Company heat-collector units like this may eventually be used in groups for heating purposes. Collector tubes utilize both direct and reflected solar energy.

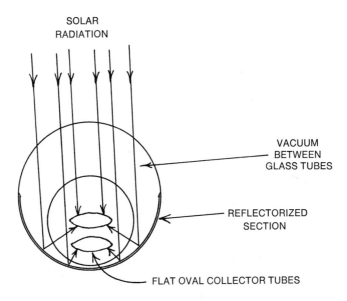

SOLAR
RADIATION

VACUUM
BETWEEN
GLASS TUBES

REFLECTORIZED
SECTION

FLAT OVAL COLLECTOR TUBES

How Corning unit works. Space between inner and outer glass tubes is evacuated to provide high degree of insulation, while tubes act as double heat trap. Black elliptical metal tubes receive sun heat from above directly, from below by reflection, so fluid is heated during entire flow circuit.

vacuum provides insulation of a quality comparable to that of a vacuum bottle. A mirrored inner surface on the back of the outer glass tube reflects sunlight onto the back of a blackened, flat-sided metal tube inside the inner glass tube, while direct sunlight strikes the front surface of the metal tube. Thus, the tube is sun-heated all the way around, and protected from heat loss by a vacuum. As with other unit-type collectors, grouping is necessary to attain a required output.

Air heat collectors

Using air instead of water or antifreeze as a heat-transport medium, this type of sun-heat collector, in its usual form, consists of overlapping black-painted aluminum louvers exposed to the sun under a glass cover. The louvers are so arranged, as shown in the drawing, that one-third of each one is in sunlight, the other two-thirds shaded by the louver above it. This provides the necessary heat-dissipation area for the sun-heated portions of the louvers. Such units may operate by natural convection or, where they must be located above the area to be heated, by blower-driven air. As the specific heat of air is only .24, compared to 1 for water, a large volume of air must be moved to do a given heating job. (The relationship of the specific-heat figures indicates that weight for weight, water can carry about four times as much heat as air can carry. And, of course, you need a very large volume of air to match the weight of the water in a fluid heat collector.) Advantages of air as a transport medium are that it can't freeze, and that a small amount of leakage has far less serious consequences than water leakage.

Other air-type heat collectors utilize a variety of materials, ranging from gauze to glass. Where glass is used, it is in the form of louvers, part clear, part black. The gauze, dyed black, is used in "matrix"-type air collectors, in which air is passed through the sun-heated gauze under a glass cover. Still other forms use finned metal plates or V-corrugated sheet metal as the heat-absorbing surface. The finned metal plates are mounted under glass with the flat side up, painted flat black. The sun heats the plate, which transfers its heat

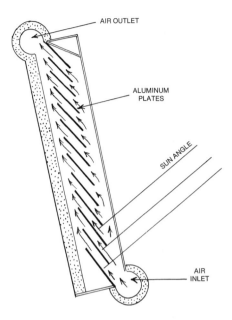

Air-type solar-heat collector. Angled louverlike plates are of aluminum arranged so that one-third of each plate is exposed to sun, two-thirds shaded by plate above. This is an efficient proportion for heat-absorbing and heat-radiating areas. Unit can operate by natural circulation, with cool room air (from floor level) entering bottom of unit by duct through house wall, sun-heated air from top of unit passing back into room through upper duct near ceiling. Unit should be on south wall or south slope of roof. If on roof, a blower must be used to move air through heater and ducts.

through the fins on its underside to air flowing through a passage beneath the plate. The same general principle is used with the V-corrugated sheet-metal types.

Sun-tracking flat-plate collectors

This arrangement, similar in purpose to that used with concentrating collectors, is designed to rotate the collector around one or more axes in order to receive a higher percentage of the available solar radiation throughout the day. The plate may simply rotate around a

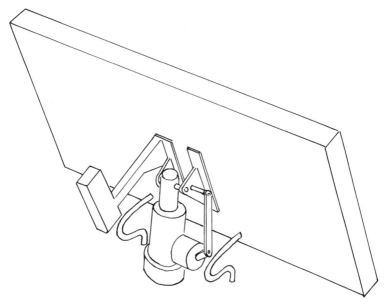

Flat-plate collectors of fluid or air type may be mounted on timed sun-following mechanism for day-long efficiency, but cost of installation is much higher than for stationary collectors.

vertical axis to follow the east-west motion of the sun; or it may rotate on a horizontal axis to follow the sun's ascent and descent; or it may rotate around both axes. The third system can sometimes increase the collector's total output by more than 50 percent. Because of the added initial cost and maintenance of the mechanism, however, it is usually more economical to simply increase the area of a fixed collector by 50 percent to accomplish the same increase in output.

Skylight sun heating

Skylights, like south windows, can provide an appreciable amount of sun heat. But, like the windows, they require insulation at night and in cloudy weather to prevent heat losses that might counteract their heat gains from sunlight. Because of their ceiling location, in fact, skylights would be likely to lose an even greater amount of heat than windows, as inside temperatures near the ceiling are usually higher than the room average. Thus, the inside-outside temperature difference and the rate of heat transfer are greater. To permit skylight sun heating with minimal nighttime heat loss, Skylid insulated, automatic skylight louvers are manufactured in sizes from 46 × 50 inches to 68 × 122 inches. The units are mounted under conventional glass

Skylid solar-heating system, developed by Zomeworks Corporation of Albuquerque, opens automatically to admit sun heat through skylight or large south window area, closes automatically when sun sets and during periods of heavy overcast. Mechanism is simple. Freon, moving from sun-warmed container on outside of control louver to cooler container on opposite edge on inside of same louver, tips the balance that opens louvers. Action is reversed when outside temperature drops below inside temperature. Units are available.

skylights, which may be either horizontal or pitched. Although the louvers open automatically during sunny periods and close automatically during very cloudy periods and at night, no electricity is required for their operation. The activating mechanism consists simply of two canisters of Freon, one on the outside surface of the louver, the other on the inside surface, with a connecting tube between them. When the sun is shining, its heat expands the Freon in the outer canister, causing it to pass through the connecting tube to the inner canister. The shift in weight opens the louvers. At night, or during cloudy periods when the inside canister is warmer than the outer one, the process reverses, and the shift in weight closes the louvers. A manual override lets you open or close the louvers or adjust the degree of opening at any time regardless of the automatically set position. The Skylid louvers are manufactured by the Zomeworks Corporation, P.O. Box 712, Albuquerque, New Mexico 87103.

Sun heating with barrels of water

As shown in the photograph, 55-gallon metal drums can be stacked in a rack inside a large south-window area to provide both heating and stored heat. The ends of the drums facing the window are painted flat black to absorb sun heat for storage in the water the drums contain. Once the drums are fully heated, of course, they also radiate heat to warm the room. Insulating doors outside the window area are opened on sunny mornings and closed at night and during cloudy weather. The metal drums are readily available secondhand from scrap-metal dealers and other sources. The racks for supporting them are made by the Zomeworks Corporation, mentioned above.

Sun-heated water from plastic tanks

One of the simplest solar water heaters to install and connect is the Hitachi Hi-Heater. Made of flexible waterproof black plastic, it

Drum-wall solar heater by Zomeworks. Bank of 55-gallon water-filled drums is mounted on rack behind south window wall. Outer ends of drums are painted flat black to absorb heat from winter sun, store it, and radiate it inside house. Insulating doors are moved between the window wall and the drums at night to retain the stored heat. Doors remain closed in summer. System may also be used with overhang to block summer sun. Racks for drums are manufactured by Zomeworks, and are available in several sizes.

serves as both heat collector and hot-water storage tank, and can be connected with flexible or rigid plastic pipe. The capacity of the model shown in the photograph is 44 gallons. When used in areas where winter temperatures fall to freezing, the unit must be drained at night and during sunless periods. An insulated inside tank may be provided to store the drained hot water for use at night. More than 200,000 of the units are already in use in Japan and the Far East. The heater is manufactured in Tokyo and is available through Hitachi America, Ltd., 437 Madison Avenue, New York, N.Y. 10022.

Hitachi solar water heater shown here is sold ready-made as a unit for mounting on house roof. Made of special plastic, it is connected to water system with plastic tubing. In areas where freezing is likely, it must be drained on winter nights.

Selecting sun-heating equipment

Before planning an installation for space or water heating, it pays to spend some time thinking in sun-heat terms. If the system is to be installed in an existing house, look for possible locations for collectors, and for the paths the piping from them could take with the least difficulty. Consider all the features of the house, and how they might fit into an overall sun-heating plan. If there's a broad south-facing wall, for example, give thought to the possibility of a large window area in it. With insulating panels, this could supply supplementary heat. Perhaps a narrow projecting roof could be installed above it to serve as the overhang to shade the window area in summer. (Information on planning overhang dimensions is given earlier in this chapter.) If the window area heats a major room or portion of the house during sunny days, the chances are the fuel saving will be worthwhile. But don't forget the insulating panels.

If you plan in terms of fluid heat collectors and heat storage, look for a space to put the storage tank. The path of the plumbing from the collector to the tank should be as direct as possible. If the basement is finished and used as a playroom, there may be room in an attached garage for the tank, but it must be well insulated. If existing space is very limited, an insulated tank room can be added without undue expense. The usual heat-storage tank requires a floor space only about 10 feet long and 5 feet wide. But allow for making the connections and, most important, getting the tank into the room. It's wise to allow walking space around the tank to provide for modifications in the system or possible repairs.

Compare the possible systems, the work involved, and the job you want done. Look for space in which the ducts might be placed. Where basement ceiling joists are still exposed, as in unfinished rooms, rectangular ducts can be fitted between joists. If central air conditioning ducts are already in place, look for a logical point where a fan-coil unit might feed warm sun-heated air into them. If you're in doubt, check with the air conditioning installer. Many of them are familiar with sun heating, and may be able to offer constructive suggestions.

If you're planning to build a new house or are already in the process, you can probably incorporate sun heating more easily. Should your building budget not allow for the complete job, you can plan for it. Providing space for a sun-heating system to be added later will save work and cut costs when the time comes. And when it does, check again with the manufacturers listed in this book. New sun-heating products and components are now in development. They may be ready when you are, and they may make your job easier and better. Follow the requirements for use of patented systems.

4 The *R* Value—How to Hold On to Stored Heat

When you build a solar-heating system you must make certain that your house retains the heat as well as it possibly can. The reason: solar-heating systems must collect their heat from the sun during the daylight hours of fairly sunny days, and store enough of that heat to carry over through the nighttime hours and occasional periods of sunless weather. To do this at a reasonable initial cost, the building must minimize heat waste, so that the usual 1,500-gallon tank of water can store the heat required for the carry-over periods.

The arithmetic of this is simple. What technical people call the "specific heat" of water is 1.0. This means that it takes 1 Btu to raise the temperature of 1 pound of water 1 degree F. So, as water weighs 62.4 pounds per cubic foot, a single cubic foot of it can store 624 Btu when you raise the temperature of that cubic foot 10 degrees F. Since the usual 1,500-gallon heat-storage tank incorporated into modern solar-heating systems holds a little more than 6 tons of water (around 12,500 pounds), it can store almost 626,000 Btu if its temperature is raised 50 degrees F, which is well within the capability of a typical solar-heating system, given an adequate heat build-up period.

To hold that heat in the tank until it is distributed through the house (as during nighttime hours) requires adequate tank insulation, and to hold it in the house as long as possible after distribution requires adequate house insulation. So you need to know in terms of Btu just how fast heat will leak away through specific types of insulation. Then you know how quickly you have to provide heat to stay ahead of the game. In a way, it's like pumping up a leaky tire. But in heating, if the heat leaks away slowly, you need only "pump" more in at long intervals; so the heat your solar-heating system has stored in its tank can carry you over several days. But if the heat leaks away rapidly, your stored heat may not even keep your house warm overnight. That's where the *R* value of your insulation comes into the picture. That is the rating used by insulation manufacturers to compare the different insulating values of their products. More about it shortly.

Insulation measurement—the *k* value

When insulating materials are tested to determine their effectiveness, one basic measurement is based on the number of Btu that pass through a square foot of *1-inch-thick* material each hour, when the temperature on one side is 1 degree lower than that on the other side. This is called the *k* value, which is the designation for "thermal conductivity."

The *C* value

This is often referred to as the "conductance" of an insulating material, and indicates the number of Btu per hour that will pass through 1 square foot of it in *the thickness to be used,* when the temperature difference between one side and the other is 1 degree F. In other words, this is the value that applies to insulating materials in the forms you are likely to be using.

The *R* value

This is the term that tells the important fact you need to know—the *resistance* of the particular insulation to the passage of heat. It is the "reciprocal" of the *k* or *C* value. In mathematical terms that simply means that you get the *R* value by placing the number 1 above the *k* or *C* value to make a fraction. For example, if the *C* value is 5, the R value would be ⅕. Customarily, to simplify calculations, the fraction is converted to a decimal—in the example above, 0.20.

The U value

This is the big one—the total of all the heat resistances of the materials in the outside walls and other inside-outside barriers of your house shell. To get the U value you simply place the figure 1 over the total resistance. If, for example, the resistances of all the materials in a wall add up to 5, the U value would be ⅕. Converted to a decimal (in this case 0.20), it tells you how many Btu per hour *per square foot* are leaking through the wall or other inside-outside barrier. To find the actual number of Btu leaking away through *entire walls* or other parts of the house shell, simply multiply the square-foot U value by the number of square feet in the walls or other areas to be figured. If, as above, you are losing 0.20 Btu per square foot per hour through 500 square feet of wall, you multiply 500 by 0.20. And you find you are losing 100 Btu per hour through the total wall area for *each degree* of inside-outside temperature difference.

An example: How much heat leaks through your walls?

The materials with which your house is built and insulated tell the heat-loss story. And you get a little bonus from the air itself. The air film next to the inside surface of a house wall, for example, assuming still air, has an R value of .68. A similar air film next to the outside wall surface, assuming a breeze of 15 miles an hour, has an R value of .17. The square footage of the inside and outside air films is equal to the inside and outside wall areas. On inside horizontal air films with heat flow downward (as on floors), the R value of a nonreflective surface is about .92; for a reflective surface (aluminum foil), about 4.55. With heat flow upward, as at ceilings, the nonreflective figure is .61, the reflective figure ·1.32.

You may find variations from these figures, as different sources base their data on different independent tests. Only a small fraction of your heat loss is through floors. And an air space of ¾ inch or more inside the wall has an *R* value of .97. (Interestingly, this value doesn't increase noticeably even when the inner-wall air space increases from ¾ inch to as much as 8 inches.) The rest of the resistances depend on the materials in the wall, as in this example listing materials from the inside wall outward:

	R value
Inside air surface	.68
Gypsum board inner wall (½ inch)	.45
Air space inside wall	.97
Insulation board sheathing (25/32 inch)	2.06
Wood siding (bevel)	.81
Outside air surface (15-mph wind)	.17
The overall resistance *R* is	5.14

To convert the overall resistance to the overall heat transmission, or *U* value, we simply place a 1 above it, divide it out, and get .19 ($U = \frac{1}{R} = \frac{1}{5.14} = 0.19$).

The *U* value above, however, is for just 1 square foot of wall area with a temperature difference of only 1 degree F between the inside and the outside. It tells us that a single square foot of our wall will lose .19 Btu each hour at that 1-degree difference. But let's assume, for easy figuring, that the wall area involved is 1,000 square feet. Over that area we will lose 190 Btu each hour. If we further assume logical winter weather conditions, with an outside temperature of 25 degrees and an inside temperature of 70 degrees, we have an inside-outside difference of 45 degrees. So we multiply our 190 Btu (the 1-degree figure) by 45, and find that our wall, on a typical winter day, will lose 8,550 Btu each hour. If the heating season lasts from October to May, say a total of 5,088 hours, we can multiply that 8,550 hourly Btu loss by 5,088 hours, and we get more than 43 million Btu lost through that wall during the heating season, assuming the inside-outside temperature difference averages out at 45 degrees.

To appreciate the value of full insulation, let's assume we add 3½ inches of fiberglass insulation (the maximum that will fit in the space between studs in a typical wall) with an *R* value of 11. Now (although we have lost our inner-wall air space) the resistance of our wall materials adds up to 15.17, almost three times what it was before. And when we repeat our arithmetic we find we've saved almost 27 million Btu during the heating season.

The foregoing examples illustrate two important aspects of heating work. First, they show the general method by which you can estimate the heat loss of a new house before you build it. And second, they show the kind of increased heat saving possible with added insulation.

When computing the heat loss of a new house, of course, your figures must include all the walls, ceilings, floors and roof through which heat can escape to the outside. And the figuring must be based on the specific materials used in each one, as the construction of walls, floors and ceilings differs. (Walls, floors and ceilings between partially heated areas of the house, such as crawl spaces, unheated

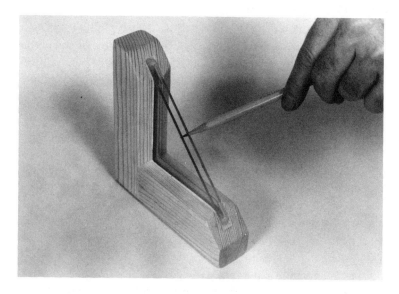

Insulating (double) glass is used in solar homes to cut heat loss. Type shown cut away can save 30 percent of heat that would be lost through single glass.

basements, etc., are figured differently, sometimes not included in the figuring at all. More about this shortly.) Window areas must also be taken into consideration, and figured according to their resistance to heat loss. Here, there's an advantage in storm windows or double glazing such as Thermopane.

Generally, following these procedures will give you a reasonably good idea of your proposed home's heat loss, though you can go into even greater detail if you're so inclined. You can, for example, obtain figures on the amount of air that actually passes *through* various types of walls, though the amount that leaks around windows and doors is likely to be a much greater factor. A $^1\!/_{16}$-inch crack around a typical front door adds up to the area of a 3 × 4-inch opening. So include weatherstripping in your plans for a solar house. Professionals also sometimes calculate the insulating effect of the house framework, such as the studs in walls and the rafters in the roof. The area they cover is sometimes figured at 15 percent of the wall or roof area.

Stripes of snow running from eaves to peak of this roof show added insulation provided by rafters. Small amount of heat leaking through insulation in attic floor has melted snow between rafters. You can see examples like this just after light snow when sun follows snow.

Where a high-R-value insulating material is used between these framing members, it will provide more insulation per inch of thickness. Fiberglass, for example, has an R value almost three times that of fir framing lumber. In a roof above an unheated attic, however, where the insulation is in the attic floor rather than between the roof rafters, the rafters provide more insulation than the areas between them. As shown in the photograph, the insulating effect of the framing members (rafters) is often clearly apparent after a light snow.

The winter design temperature

These temperatures have been computed for the purpose of planning heating systems throughout the United States, on the basis of winter temperatures in the various areas. The design temperature for a given area, however, is not the lowest winter temperature you can expect. It is the lowest temperature encountered approximately 97.5 percent of the time. Thus, it does not include the extreme lows sometimes reached for a few hours between midnight and dawn during about ten days of the year. If you were to plan your heating system (solar or conventional) on the basis of the extreme lows, you would have an oversized system that would be far larger than necessary for all the rest of the heating season. Basing your heating system's capacity on the design temperature, therefore, keeps initial costs down, and has very little effect on your overall heating. During the short periods when the outside temperature drops to an extreme low (below the design temperature), your inside temperature may fall below your thermostat setting by several degrees, but seldom by enough to affect your comfort. Check local fuel suppliers or your nearest weather station to find the outside winter design temperature for your area.

The *R* value of the materials

The table of R values provides approximate figures for the common building and insulating materials. And your supplier can provide values for the specific insulating materials you buy. Many are labeled accordingly, as *R*-7, *R*-11, etc. Add these values to find the total thermal resistance of your house shell, and the *U* value as described earlier.

Where insulation saves the most heat

Insulation does the most for you where the most heat is found. So, since heat rises, you get the most for your insulating dollar on a per square foot basis in the ceiling or roof. That's where the biggest inside-outside temperature difference is usually found. But on an overall basis, insulation in the walls has a greater effect, simply because the total wall area is usually much greater. In a typical *uninsulated* house, tests indicate relative heat losses approximately like this:

Walls	33.0%
Roof and ceilings	22.0%
Floors	0.3%
Doors and windows	30.0%
Infiltration	14.7%
	100.0%

These losses, of course, would vary in different types of construction, but would generally retain about the same proportions.

Installing insulation

In new construction, insulation is simply put in place in the walls and ceilings before the inside wall and ceiling surfaces are applied—a relatively easy job. Fiberglass, in one of several forms, is the most widely used type. Often termed "blanket insulation" when in rolls, or "batts" in 4-foot lengths, it is available with or without a "vapor barrier." The barrier takes the form of a treated paper or metal-foil covering, which should be toward the "warm" interior side of the wall when the insulation is installed. It retards the passage of water

Fiberglass attic insulation made in widths to fit between joists can be applied over existing insulation. Thickness of 6 inches is adequate. More will cut heat loss still more.

vapor through the house wall to the outside surface, where excessive moisture from this source could cause such problems as peeling paint.

A vapor barrier is especially important in the outside walls of kitchens and bathrooms. Insulation with an integral vapor barrier is often made with the paper flanges extending from the sides so the insulation can be stapled to the framing along each edge. "Pressure-fit" insulation is simply pushed in place between the framing members (studs, etc.). It is made slightly wider than the standard stud and rafter spacing so friction (and there is plenty) holds it in place. If pressure-fit insulation without a vapor barrier is used, polyethylene sheet vapor barrier is usually applied over the inside of the framing after the insulation is tucked in place, before the inner wall surface is applied. Use a staple gun to fasten the poly sheet to the framing;

when the wallboard is nailed in place, it also holds the poly sheet.

Important: try to avoid compressing insulation. When *R*-7 insulation (about 2 inches thick) is compressed to 1¾ inches, its thermal resistance drops to about *R*-6. Equally important, if you use insulation with a reflective covering, such as aluminum foil, allow at least ¾ inch between the reflective surface and the nearest structural surface, such as the inside surface of the interior wallboard. Without this space (preferably 1½ inches if it's in a ceiling) you won't get the benefit of the heat reflection, which is appreciable. As an example, *R*-11 insulation installed in a ceiling with reflective foil side down, and a ¾-inch space between it and the adjacent structural surface, gains about two units of resistance, for a total of *R*-13. But again, the air space must be at least ¾ inch.

Foam insulation may also be put inside walls, though it is probably most widely used as "perimeter insulation" around foundations, including slab foundations. Made for the purpose, the foam insulates the foundation from the frozen ground in winter, and is practically unaffected by moisture. The typical *R* value in a 1-inch thickness is 4.

Pourable insulation in granular or pellet form is often used between joists in unfloored attics, as it can be leveled off at the desired thickness. And if additional insulation is desirable or proves necessary later, it is a simple matter to add it. Just pour in a few more bags of the granules or pellets, and level off at the increased depth.

Easy insulation for old houses

Adding to the insulation of an existing house, as may be required for the addition of effective solar heating, can be complicated unless special methods are used. Older homes, especially those built before the 1920's, may have no insulation whatever. The place to add insulation with the least work (if it hasn't been done already) is in an unfloored attic, as mentioned. (The attic is usually unfloored if it has insufficient headroom for conversion to living space.) Roll or batt insulation can be laid between the rafters. You can calculate the

Cellulose insulation made from recycled materials can be blown into existing walls from inside or outside. R value is about 3.70 per inch of thickness. Hire professionals to do it or, in some areas, rent a machine to do it yourself. Bags of insulation are emptied into machine's hopper. Machine blows it into walls through hose. For details, contact the National Cellulose Manufacturers Association, 400 West Madison Street, Chicago.

Working from inside house, first step is boring holes through wallboard of outside walls to take blowing nozzle. In rooms over unheated crawl space, bore through floor to blow in insulation.

Flexible hose carries insulation from blowing machine to pointed nozzle that blows it into walls. Insulation can be blown into ceilings, as where rafters are topped with roof decking.

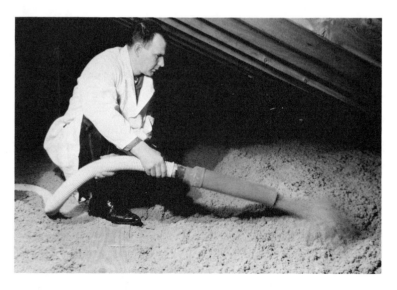

Where attic is accessible, insulation can be blown between and over joists.

reduction in heat loss as described earlier. The thicker the added insulation layer, the greater the heat saving.

If insulation must be added to finished walls and ceilings, or floors of an older home, there are several approaches. It can be blown into the air space within the walls by professionals, using special equipment. This can usually be done in about two days if the house is of average size. Look in your yellow pages under "Insulation Contractors," for firms that do this work, and get an estimate. The job is usually priced by the square foot. Typical procedure: clapboards or shingles must be removed from the top and bottom of the walls to be insulated. Then holes are cut through the sheathing in the spaces between studs to admit the hose through which insulation is blown into the walls. When the inner-wall space has been filled with insulation (usually a special form of fiberglass), the holes in the sheathing are closed and sealed, and the clapboards or shingles replaced. The insulating effect is comparable to that of full-space insulation installed during construction. It cuts conventional heating costs and can often make partial solar heating possible.

Do-it-yourself old-house insulation

If your budget is too limited to allow for professional help, you can add insulation to an old house yourself. To cut heating costs or make partial solar heating feasible, you can add the insulation to the *inside* of existing interior walls and ceilings. This is done by applying insulating foam panels to the existing walls and ceilings with an adhesive mastic made for use with the foam. A layer of wallboard or paneling is then added to cover the relatively soft foam. As typical foam has a higher thermal resistance than the usual fiberglass insulation, it need not be as thick, which makes its interior use practical. (The R value for heat resistance at 1 inch thickness may be 4.50 or more for some types of foam, compared to around 3.70 for fiberglass.)

Baseboards and wall moldings are removed before the foam is applied. The treatment of trim around doors and windows depends on the individual job. In some instances it may be removed and replaced over the foam, in others it may be allowed to remain, and modified to cover the edges of the foam and the wallboard or paneling that cover it. Base the procedure on the nature of the trim and the relative amount of work involved in either method. The foam is available through lumberyards. Verify the *R* value before you buy. Common thicknesses are 1 inch and 2 inch. Check with your local building inspector on any special requirements before you plan the job.

The methods of heat-loss figuring in the foregoing pages are not the only ones used for the purpose, though most other systems are based on the same principles. In some procedures, charts are used in which a portion of the figuring has been included. In others, allowances are made for "average" window areas or other factors to simplify your work. If you're buying a backup heating system, mail order, for example, a booklet may be available containing such a pre-figured system. If so, it will be keyed to the heating units, and is likely to save you effort. And knowing the basic principles described in this chapter will help you avoid errors.

A final suggestion on insulating for solar heating: use as much insulation as you can afford. It makes the most of solar heating and cuts the cost of operating the conventional backup system. And if you air condition, it cuts costs again and increases cooling effectiveness.

5 Solar Heat for an Existing House

Judging the feasibility of adding some form of sun heating to an existing house isn't difficult. The first step is simple observation. Take a look at your house from the outside at intervals on a sunny day, preferably during the heating season. You need to know which areas of the walls and roof are in sunlight during a major portion of the day. Be sure to note whether any of these sunlit areas are shaded by trees, adjacent houses, or anything else during any particular part of the day. If your house is so situated that no area large enough for adequate sun-heat collection is in sun throughout the middle two-thirds of the day, look for a lawn area close to the house that receives full sun during this period. Heat collectors on the lawn are often connected to a building by insulated pipes.

If any wall of the house faces due south or a little west of due south, consider the possibility of using a major portion of that wall for window heating, as described in Chapter 1. Windows of conventional size, already in a south wall, are already contributing sun heat to your house, but probably not in amounts large enough to be apparent in fuel savings. In your survey of sun-heating possibilities, keep in mind, too, the other forms shown in Chapter 3. Your house may have individual features that lend themselves especially well to particular forms of sun heating. And a combination of systems often brings the best results.

When you have decided which type or types of sun heating may do the heating job best for you, look into the insolation factor of your region. You can get a rough idea of it for heating purposes from the February insolation map in Chapter 2. If the figures look promising,

Typical raised ranch-type modern house.

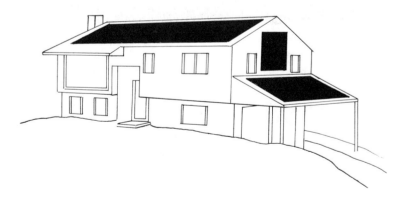

Same house, black areas indicating possible locations for solar-heat collectors, depending on orientation. If entrance faces south, entire south slope of roof can be collector. Windows at left can be enlarged, suitable overhang added. If right side of house faces south, roof may be added over garage entrance to house heat collector. Vertical heat collector may also be used on upper section of south wall.

you can obtain year-round maps from the sources mentioned. If you're thinking of future solar-powered air conditioning, the July map in the same chapter will give you a basis for estimating.

Pre-planning

The practical, mechanical part of pre-planning comes next. If you have settled on a flat-plate, fluid-type heat-collector system (often most applicable to an existing house), you'll require some basic information about your house. When the collectors are to be on the roof, you'll need to know the size and spacing of the roof rafters and the approximate weight of the heat collectors. If you intend to use ready-made collectors, you can get all the specifications regarding sizes and weights from the manufacturer. If you plan to build your own, you can figure their approximate weight as described in Chapter 6, and you can build them in sizes that best suit the situation. In most cases, conventionally sized roof rafters can support flat-plate collectors of typical design and construction. But to play safe, check with your local building inspector before the work begins. If any reinforcing or trussing is advisable, it's easier to do before the collectors are mounted than after. Whether such measures are necessary depends on the individual house. The rafters, for example, may already be close to the minimum allowable cross-sectional size for their span and spacing if the roof is a wide one. Or, in a narrower roof, they may be well above the minimum allowable cross-sectional size, and more than adequate to support the added load.

This is also the stage of pre-planning for mapping out the possible paths for the plumbing that goes with the sun-heating system. It must pass through the roof at a number of points (depending on the arrangement of individual collector sections) and it must continue down through the house to the heat-storage tank as directly as possible. If it can't run down through closets for concealment, think about hollow false posts built up from nominal 1 × 6 or 1 × 8 board-grade

Modified Cape Cod house.

If entrance faces south, roof may house heat collectors in several sections, divided by dormers, as indicated by black areas. If left end faces south, roof over garage door may be added to house heat collector, as shown. South windows may also be enlarged, overhang added. If back of house faces south, entire south slope of roof can be utilized for heat-collector area.

Similar house without dormers.

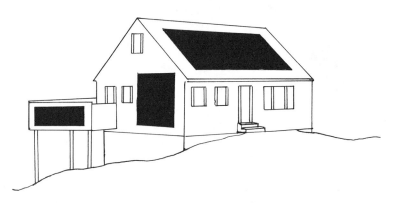

If entrance faces south, entire south slope of roof is available for heat collectors. South windows may also be enlarged, with overhang extended as necessary. If left end of house faces south, vertical heat-collector panels may be installed in house wall, also in end of porch railing wall, as shown in black. Similar use of opposite end wall, if it is south wall, is possible.

Simple one-story design with basement garage.

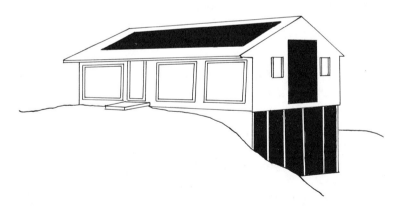

If entrance faces south, front windows may be enlarged, overhang added for window-wall type of solar heat. Entire south slope of roof may also be utilized for heat-collector area, as shown in black. If right end faces south, lower (foundation) portion may be used in Trombe-Michel manner. Vertical fluid-circulating heat collector can be used in upper wall section, as indicated. Opposite end, if facing south, can be used in same way. If rear of house faces south, the south roof slope and rear wall can house solar-heat collectors.

lumber. It's usually a lot easier to run a pipe between floors inside a hollow post against a wall than it is to run it inside the wall, whether the wall is a partition between rooms or an outside wall of the house. And if it happens to be one of the outside walls, the false post has the advantage of keeping the pipe completely *inside* the room for minimum heat loss. Of course, any loss that occurs inside the house actually contributes heat to the room in which the loss occurs. However, as the collectors are located on the south side of the house, the pipes running from them are usually in rooms on the south side. But north rooms are likely to need more heat because their shaded outside walls lose heat faster than sunny southern ones.

So it's obviously best to minimize indoor heat losses from piping so the heat can be used where it's needed most. The simplest way to keep the heat in the pipes is through the use of ready-made flexible foam pipe insulation. This is made to fit standard copper water-tube sizes. If you're using galvanized steel pipe (which has a larger outside diameter) you can usually insulate it with the next larger insulation diameter than you would use on copper tube of the same "nominal" size. In its most convenient form, the insulation resembles a thick, flexible tube, slit lengthwise from end to end. It's usually sold in 4-foot lengths. You slip the insulation over the pipe (after the pipe is installed) by opening the slit with your fingers. The slit tends to close itself as soon as the insulation is in place, but should be fastened with waterproof stick-on tape. The tape is available from the insulation supply house.

The existing plumbing

In order to map the plumbing, of course, you must have a location for the heat-storage tank and other components of the system. The best tank location is usually a heated part of the basement, if space permits. Typically, the tank is 5 feet in diameter and 10 feet long, which means that a somewhat larger floor space is required to allow for connections and insulation. (In the Thomason solar homes, the tank is uninsulated, but surrounded with rocks in an enclosing bin, as both the tank and the rocks play a major role in the radiation that warms the house.) In some older homes with deep cellars, the tank may be installed vertically (as it is in some modern solar homes), requiring only about half as much floor area.

Most important, you'll have to plan a way of getting the tank into the basement. If that can't be done without costly wall removal and possibly excavating work, there may be room for it in an attached garage. In a garage that isn't heated except by basement heat passing through an intervening masonry wall (frequently the case in new homes), the tank itself can be heavily insulated, along with all plumbing connected to it. Any fan-coil unit included in the system should be inside the heated basement.

In basementless houses, the only answer may be to add a small room to accommodate the tank. If you build such a room, insulate it to the maximum. The total area to be insulated will be relatively small, so the cost won't be excessive, and the advantages are obvious. Insulation of the tank itself depends on the individual situation. If the tank is located in an enclosure built to house it, such as a plywood bin, pour-in insulation can be used to fill the space between the tank and the bin walls. This type of insulation, often used to fill the space between joists in an unfloored attic, has the advantage of ease of installation. If the tank is located in an open area of the basement or garage, without an enclosure, it can be wrapped in thick fiberglass insulation with paper or foil backing on the outside, held in place with stick-on tape. One final caution regarding tank installation: provide very solid support for the entire tank, as the usual 1,500-gallon size weighs more than 6 tons when filled.

If the system requires ducts, they can usually take the most direct route through the basement, with main ducts between the basement ceiling joists. In a house with no basement, pipes can be run from the ground-level tank to a fan-coil unit in the attic. Ducts from there can be run to the rooms. The important point is advance planning in detail. Many houses are successfully heated by conventional warm-air heating systems ducted through the attic and emitting warm air through ceiling or high wall registers.

If the house is presently heated by a hot-water radiant system with heating pipes embedded in a concrete slab floor, the sun-heating system can usually be interconnected to provide heat through the same system while retaining the conventional boiler as the backup heater. Because of the possible variations in systems, it's usually best (unless you're familiar with heating work) to have this part of the job done by a heating installer—if possible, the one who did the original work. If he is familiar with sun heating, as many heating installers now are, he may also be able to take on other parts of the job.

Wall-mounted collectors

Where fluid-type heat collectors are mounted on the outside walls instead of the roof (as when the ridge direction makes roof mounting difficult or complicated), they should be attached so as to provide full support without imposing any load on the pipes connected to them. In general, the collector area should be somewhat greater than for collectors mounted with their surface at right angles to the sun, as described in Chapter 2. Where wall collectors are mounted *above* a low-pitched roof, such as a porch roof, their efficiency can be boosted by placing a large-area reflecting surface on the roof directly in front of the collectors. Aluminum sheet or aluminum-asphalt roof coating can be used. Sunlight reflected from this surface to the collector provides added input at minimum cost. Also consider the possibility of building a narrow overhang above the collectors to shade them in summer and avoid overheating problems if you have no use for sun heat in hot weather.

All piping to and from the wall collectors should be insulated as usual with the pipe insulation described earlier. If you use another type of foam insulation on any portion of the piping, be sure it can withstand the temperature. Some foams, intended for lower-temperature insulating applications, may actually soften and collapse at the temperatures common to fluid-type flat-plate collectors.

In the event that window-type sun heating suits the situation, to augment either conventional heating or another form of sun heating, you'll have to plan adequate support for the structure above the window area. This requires doubled beams (called lintels or headers) of nominal 2-inch thickness across the top of the window area. Your local building code will specify the cross-sectional beam size for the span. Typically, you'll need doubled 2 × 8's for spans of 6 to 8 feet, 2 × 10's for spans of 8 to 10 feet. For the large areas needed for effective sun heating by this method, it's best to decide first on the window sizes you'll use to make up the overall area, then space the supporting posts to accommodate them. The span between posts will determine the size of the doubled beams above the windows. Windows of this size, of course, should be of double-pane insulating glass. And to minimize nighttime heat loss, they should be fitted with insulating foam panels for use at night. You can make these as described in Chapter 6.

The existing heating system

If your present heating system can be left intact, you'll be assured of the same heating performance you've always had, regardless of weather and the heat contribution of your sun-heating system. The conventional system's thermostat has only to be set to switch on at a temperature a little below the switch-on temperature of the sun-heat system, and it won't run as long as the sun system can maintain its temperature. When the sun system can't keep the temperature at the set level, the conventional system will switch on as soon as the temperature reaches its lower switch-on setting. A little experimenting may be necessary to get the setting right. So long as the temperature in the heat collector is higher than that in the tank, the sun system will continue to store heat. When the collector is cooler than the tank, it will shut off.

Whatever heat the sun system contributes to the heating job reduces your fuel bill.

Insulation

Adequate insulation added to an old house can also make sun heating more effective. If the present insulation is scanty or absent altogether, you can use one of the methods described in Chapter 4 to bring it up to par. In some areas you can rent equipment to use yourself for blowing insulation into walls. For the nearest source of cellulose insulation for this purpose, write to the National Cellulose Insulation Manufacturers Association, 400 West Madison Street, Chicago, Illinois 60606. To find professionals to install blown-in fiberglass and other types of insulation, look in the yellow pages of your local phone book under "Insulation Contractors."

Small-scale sun heating

If you'd like to attempt sun heating on a small scale before you tackle a major installation, you might try it on domestic water heating, as described in Chapter 3. Other possibilities might be small-area space heating, as in an enclosed breezeway or an unattached garage. If the area has no present heat, some form of thermostatically controlled backup heater, such as an electric one, is needed in winter-freeze regions to prevent freezing of the heat-storage tank when the sun system is inoperative in cloudy weather or on very cold nights. In such small areas a water tank of around 80 gallons capacity could be used uninsulated to serve both for heat storage and for radiation. Protect the collector with antifreeze pump-circulated through a ready-made domestic hot-water-type heat exchanger in the tank. A thermostatic shut-off should be provided to stop the pump when the collector cools. This, of course, should be considered experimental sun heating. The aim should be to keep a small area comfortable without fuel in moderate winter weather and, with adequate backup, above freezing in colder weather—an aid to car starting in an other-wise frigid garage. As with any heating system, adequate insulation is essential.

6 How to Build a Flat-Plate Solar-Heating System

Whether you plan your own solar heating or work from an engineer's plans, you can cut costs by building the system yourself. You may also be able to save by careful shopping for major components of your system, some of which (such as fan-coil units and storage tanks) may be available from building wreckers. If not, they can be bought new from easily located sources, as described later. The basic plumbing of the system requires only standard materials sold by plumbing suppliers.

You can also cut costs by improvising. The fan-coil unit, for example, really consists of a radiator (much like an automobile radiator but larger) and a squirrel-cage blower of the type used in conventional warm-air heating systems. If you are an experienced do-it-yourselfer, there's a good chance you can devise this piece of equipment for a lot less than the cost of a readymade unit.

The components of a good-sized professionally built solar heating system and a small-sized home-built experimental one are shown in photos later in this chapter. The smaller unit is an example of what you might try as a first effort. Later, if its performance comes up to your expectations, the larger system, perhaps including readymade collectors, might be your next goal.

If you build your own heat collectors, they are most easily made from stock-size copper tubing and fittings from your plumbing supplier, and copper sheet of the thickness specified later, obtainable from a metal-supply house. Two metal-supply firms are: Metal Goods Corporation, 8800 Page Boulevard, St. Louis, Missouri 63114; Joseph T. Ryerson & Son, Inc., 16th and Rockwell Streets

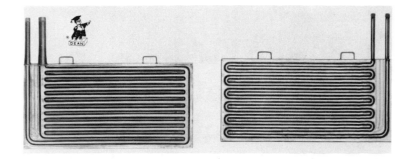

Another type of heat exchanger that can be used in solar heating. These panels, made by Dean Products, Inc., are formed from flat metal to the shapes illustrated. Type at left is called parallel flow type, as fluid passes along numerous parallel passages between inlet and outlet. Type at right is called series flow, as fluid passes along series of continuous hairpin turns in single passage from inlet to outlet. First type is suited to solar-heat collector. Second type can be used as heat exchanger in tank.

(mailing address: Box 8000A), Chicago, Illinois 60680. You will probably find local branches of these or other metal-supply companies in the yellow pages of your phone book. There is usually a minimum charge for material to be delivered, but the cost of material for a solar-heating system is likely to equal or exceed it. The rest of the materials for the home-built collectors shown in the plans are stock items from lumberyards, hardware stores, and glass dealers.

If you'd rather avoid the work of building your heat collectors, you can buy the ready-made types shown, from the manufacturers mentioned, or others. Dean panel coil units have been used as industrial heat exchangers for many years, and are also suited to solar heat applications. Dean Products, Inc., is located at 985 Dean Street, Brooklyn, N.Y. 11238. Sunworks, Inc., another reputable manufacturer, is located at 669 Boston Post Road, Guilford, Conn. 06437. Or you can buy your collectors in kit form, like the Revere units shown in the photos and drawings. Details on these kits can be gotten from Revere Copper and Brass, Inc., P.O. Box 151, Rome, N.Y. 13440. The panels used in this system are based on the Revere laminated copper building panels. These consist of a substrate, usually exterior-

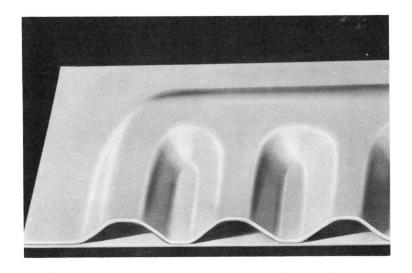

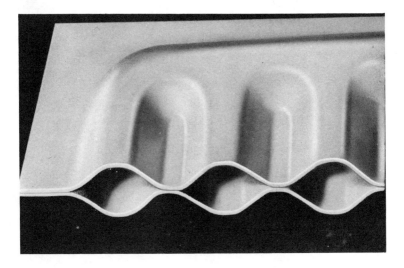

Panel that is flat on one side is called single embossed. When both sides of the panel are formed as shown, the panel is called double embossed.

REVERE COMBINATION LAMINATED PANEL ROOF AND SOLAR COLLECTOR

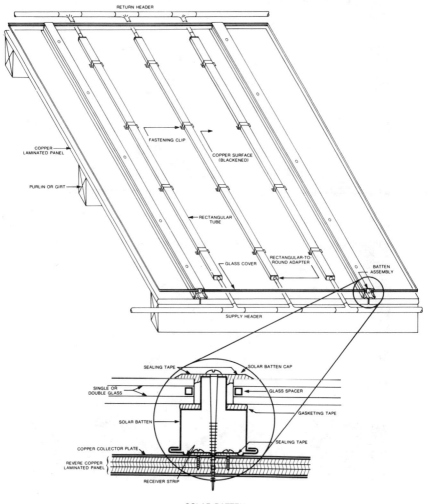

How Revere solar-energy collector panel is laid out. Detail of batten that joins panels is at bottom. All parts are keyed to system. Insulation is used under panels. Headers at top and bottom are boxed in to suit roof design, and packed in insulation.

Revere system test panel. Glass is being fitted into grooves made to take it. Panel tubes run horizontally in test panel, but run vertically in actual use.

grade plywood with a layer of copper factory-laminated to what will be the upper surface, and an aluminum balancing sheet laminated to the other surface to prevent unequal absorption of water vapor. The copper heat-collector tubes used with the panels are rectangular rather than round in cross section, so they can be attached broad side down to the copper roof layer (which serves as the heat-collector plate) with clips instead of solder. The broad face of the rectangular tube, of course, provides more heat-conducting area between plate and tube than is possible with round tubing on a flat plate, unless the plate is formed to provide more contact surface between it and a round tube. A special heat-conducting epoxy containing copper particles is also used to prevent formation of deposits between the rectangular tubes and the plate, and to further enhance heat transfer between the tubes and the plate. All materials necessary to assemble the heat collectors are available from the manufacturer. These include adapter fittings for connecting the rectangular tubes to standard round tubing, fittings for connecting rectangular tube ends

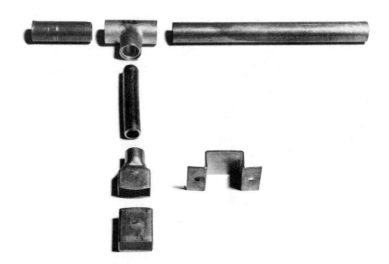

Fittings for use with Revere rectangular-tube solar-energy collector system. Parts shown separated. Top is ¾-inch header with T and reducer bushing to connect to nominal ⅜-inch collector tube. This tube connects to the special fitting that takes rectangular tube. Bottom: Coupling for rectangular tube. At right: Clip that holds rectangular tubes to copper-laminated plywood roof panels.

to increase length, solar battens that can support either single or double glass over the heat collector, clips for the tubes, heat-conducting epoxy, sealing tape, and heat-absorbing black paint. The paint used is Black Velvet, manufactured by the 3-M Company.

Two other forms of Revere heat collectors are available. One is a completely assembled collector unit, including single- or double-glass covers, ready to connect to the system's plumbing. The other is a collector designed specially for heating swimming pools to extend the swimming season. The panel size of the field-constructed unit is 2 feet × 8 feet, the completely assembled unit is 3 feet × 6½ feet, and the pool-heater model measures 34 inches × 74 inches, including headers. For large jobs, other sizes may be obtained on special order. Units are combined to make up the required total area.

Other materials

Once you have decided on the heat-collector type (ready-made, kit, or home-built) and the area, you can shop for prices on the other materials the system requires. These include plumbing, controls, tank, fan coil, and ducts. If you live in a mild climate, your initial costs will be lower. You're likely to need less collector area, and you may get by with a 1,000- instead of a 1,500-gallon storage tank. If prices are close, favor the larger size. Specify the location of fittings you want on the tank to suit your system, and a manhole with a bolt-on cover. The manhole is required in order to assemble the main heat exchanger inside the tank. (You don't need this in areas where temperatures remain above freezing.) The smaller heat exchanger for domestic hot water can be installed from the outside. In many localities, tanks of this size are usually built to order by sheet-metal or welding shops, so individual specifications are the rule rather than the exception. If, however, you discover a tank that approximates your needs at a building wrecker's yard, and it's in good condition, you may find it can be altered to your requirements by a competent welder for a lot less than the cost of a new one. The best approach: get new-tank prices from several suppliers, and check used-tank possibilities with nearby building wreckers. (Used tanks may not be available, or they may be in poor condition. But the possibility of finding a good one is always worth checking.)

Building codes

Before actually buying your materials, check your local building code in regard to the roof structure on which the heat collectors will be mounted. And compute the total weight of the collector, including all parts—copper sheet, tubes, fittings, glass, framework, and the weight of the water in the tubes. Copper sheet of .016-inch thickness (nearest B&S gauge is 26) weighs ¾ pound per square foot. (This is a thickness widely used in fluid-circulating flat-plate collectors.) So a collector with 400 square feet of area would have a sheet-copper weight of 300 pounds. If it contains 600 feet of nominal ⅜-inch Type L copper tubing (the type commonly recommended), weighing .198 pound per foot, the total tube weight, not including headers, would

COPPER TUBE SIZES

Nominal tube diameter, or "size" (inches)	Actual outside tube diameter (inches)	Type L		Type M	
		Wall thickness (inches)	Weight (lb/ft)	Wall thickness (inches)	Weight (lb/ft)
⅜	0.500	0.035	0.198	0.025	0.145
½	0.625	0.040	0.285	0.028	0.204
⅝	0.750	0.042	0.362	———	———
¾	0.875	0.045	0.455	0.032	0.328
1	1.125	0.050	0.655	0.035	0.465
1½	1.625	0.060	1.14	0.049	0.940
2	2.125	0.070	1.75	0.058	1.46
2½	2.625	0.080	2.48	0.065	2.03
3	3.125	0.090	3.33	0.072	2.68

SOURCE: Copper Development Association, Inc.

Nominal tube sizes, actual outside diameters, and weights of copper tube. You can figure inside diameters by subtracting twice the wall thickness from the outside diameter.

Weight per square foot (ounces)	Thickness (inches)
8	0.0108
10	0.0135
14	0.0189
16	0.0216
20	0.0270
24	0.0323
32	0.0431

Use these weights per square foot of copper sheet in figuring the overall weight of your flat plate collector.

be 118.8 pounds. If you use 1-inch nominal-diameter headers, you can figure their weight on the basis of .655 pound per foot. Weights of other diameters that may be used in the plumbing and supported by the roof structure are given in the chart. (If more than one collector unit is used, tube or pipe of ¾-inch diameter or larger may be used as the supply and return lines to the separate collectors.)

To estimate lumber weight, you can figure an average fir 2 × 12 weighs around 4¼ pounds per running foot, so smaller sizes are easily computed. If you use single-weight window glass, allow around 1¼ pounds per square foot. For double-weight, allow a little over 1½ pounds per square foot. Other materials used in the collector, such as the H-moldings, are best weighed during construction, a procedure that can also be followed throughout if you want total accuracy.

As to the weight of the water, simply fill a known length of the pipe involved (with one end capped or corked), then pour the water out into a container, weigh it, and subtract the weight of the container. Then divide to get the weight per foot of pipe.

From your weight calculations you'll see why heat collectors are usually built in separate sections that can be connected after lifting into place. The total weight of a collector of a typical home-heating area, however, seldom adds a per-square-foot load on the roof as great as that of a heavy snow. But as the roof may have to support the collector plus the snow, some reinforcement, such as heavier rafters (in new construction), may be advisable beyond the normal code specifications. Extra rafters, or trussing, may be used to reinforce an existing roof. Your best bet: ask your local building inspector to recommend the method. He'll have to approve it later, so this simplifies matters.

If your metal supplier designates sheet thickness by weight per square foot in ounces, such as 10 ounce, 16 ounce, etc., you can figure the weight of your collector-plate area directly. The table indicates the thickness of the weights commonly used. Although sheet sizes as large as 36 inches × 96 inches, and even 36 inches × 120 inches, are manufactured in the 16-ounce thickness, and 30 inches × 96 inches in the lighter grades, these sizes are not always available at all metal-supply outlets. If not, inquire as to what quantity you would have to buy to get the size you want on special order. In some cases a minimum charge of a certain amount (which varies with the firm) is the determining factor. If the cost of the amount of material you require comes up to the minimum charge, you can have your special order. Check on the pricing system, too. Usually the cost of copper per pound varies with the amount you buy, as in amounts up to 500 pounds, 500 to 1,000 pounds, and so on. Strip copper (usual maximum width 12 inches) can also be used if the sheet size you want isn't available. The strips may be run across the slope of the collector, overlapped about half an inch in shingle fashion. Or they may be run parallel to the slope, and overlapped by the same amount. If they are run this way, tubes should not be located close to an edge unless the overlapped edges are soldered to assure conduction of heat to the tube from both sides. Soldering of seams is worthwhile in any event to provide a waterproof surface in case the covering glass should be broken by hail or by other causes.

The glass

Double- or single-thickness window glass (also called double or single weight) may be used for the collector's glazing. To avoid problems in handling the glass and also limit repair costs in case of breakage by hail, etc., it's wise to limit the size of the panes to about 24 inches × 72 inches. If you build your collectors in sections of this size or multiples of it (to be connected after placing), the overall job will be simplified and you'll avoid the problem of shadows cast over the collector area by the glass-supporting framework.

Where there is much likelihood of glass breakage, as from hail, falling ice, or other causes, the glass can be protected by ½-inch galvanized wire mesh supported on a frame about 3 inches above the glass. This has the effect of reducing the effective collector area by around 15 percent, so if mesh is used, the collector area should be increased accordingly. For added durability use double-weight glass.

The frames in which the glass is set, as shown in the drawing, may be of wood or metal. The units shown have wooden outer frames and metal H-molding where it is necessary to join panes of glass edge to edge to keep pane sizes within reasonable limits. You can buy the H-molding from large glass suppliers, along with the sealant used to waterproof the joint. Where the glass rests in the recesses of the wooden perimeter frame, it can be sealed with a specially made sealing compound, sold by glass dealers. This is likely to have a silicone base. Sold in cartridges that fit the usual hand-operated caulking gun, it's easy to apply in the long seams common to heat collectors. It forms a strong waterproof bond between glass and frame, but in case of glass breakage it can be cut away with a sharp knife or razor to release the broken glass. The frame should be thoroughly painted with a good oil-base exterior gloss enamel before the glass is set in place.

Collector soldering

As the tubes in the collector are connected to each other and to the sheet-copper plate by soldering, it's important to select the correct type of solder. The connections between tubes (as from small tubes to headers) may be made either with standard "sweat" fittings (also called capillary fittings) or with CxM (copper to male thread) fittings, as shown in the drawing, if the headers are large enough. Connections in other parts of the plumbing leading to and from the collectors are made with standard sweat fittings. In all connections with these types of fittings a 50–50 lead-tin solder (sometimes called 50A) should be used. This contains 50 percent tin, which results in complete melting at a relatively low soldering temperature—necessary for thorough "wetting" of the copper and easy flow by capillary action into the small space between tube and fitting. The same type of solder should also be used in seams between sheets of copper, if the sheets are held tightly together.

For soldering the tubes to the sheets, however, a 40–60 (40A)

Tools for making sweated (soldered) copper-tube connections in heat collector and related plumbing. Top: Propane torch. Bottom, left to right: Tube cutter with built-in reamer. After cutting tube, use pointed black reamer to remove burrs from inside of cut tube end. Paste flux in can. Buy amount according to number of connections to be made. Wire solder. Type best suited to plumbing is called 50-50, or 50A. Buy it without flux core.

solder has advantages. This contains 40 percent tin and 60 percent lead, and though it doesn't wet out and flow as well as the 50A type, it remains "pasty" to a higher temperature. This enables it to form a "fillet" between the tube and the sheet on each side, providing better heat conduction between sheet and tube. Solders of this type are often called "wiping" solders because of their pasty nature when heated, which allows them to be spread somewhat like butter. A small wad of cloth (with plenty of layers) soaked in tallow is commonly used to wipe the pasty molten solder into grooves or spaces to be filled. Do not, however, use a solder with less than 40 percent tin, as you'll have a difficult time getting a good bond with it. If you try this wiping technique in tube or sheet soldering, use care to avoid burns, and try it first on a small part of the work.

The handiest solder for tubing connections is the round ⅛-inch

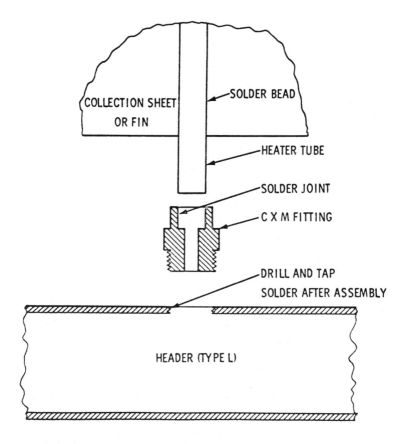

Instead of T fittings you can use CxM fittings to join heat-collector tubes, if this cuts costs in tube sizes involved. The CxM is designed to be sweat-soldered at one end (to the small tube, in this instance) and threaded at the other end (to the large header tube, as shown here). Flux the threads so the threaded connection can be soldered after assembly. The CxM fitting is normally used to connect soldered copper plumbing tube to threaded pipe.

wire type *without* a flux core. You buy the flux separately, as detailed shortly. For soldering the tubes to the sheet, which requires solder in greater amounts, twist three or four solder wires together to form a "rope." Use the correct type of solder in each instance.

The flux

Soldering flux must be used in making all the seams and connections in your collector and its related plumbing (if of copper). It performs a number of functions without which the soldering job would be difficult if not impossible. Even if the surface of the copper is perfectly clean, as after fine sanding, it will acquire a layer of oxide by the time it is brought to soldering temperature, and this oxide would make it very difficult to get the solder to adhere. The flux not only protects the copper surface from the oxygen in the air (which creates the oxide), but also dissolves oxides that may still be on the metal. Thus, it enables the solder to contact the bare metal directly, wetting it like paint. If you're unfamiliar with soldering, you can demonstrate this to yourself by applying molten solder to a piece of scrap copper with and without flux.

A number of flux types are suited to copper soldering. Paste fluxes made of petroleum jelly and zinc chloride, or ammonium chloride plus zinc chloride, are inexpensive and effective. And paste-type flux is best for plumbing work. For the sheet work you can use either a paste flux or a water-based zinc chloride flux, or a water-based one containing both zinc and ammonium chloride. To be sure of the right type for the job, buy it from the plumbing supplier where you buy your tubing, and explain what you want to use it for. You'll probably need about a pound of paste flux or a pint of water-based flux for each 8 pounds of solder. As to the choice of paste or water-based flux for the sheet-to-tube work, it's a matter of personal choice. You may find it easier to spread the liquid form (a toothbrush is often used). Wear protective glasses or goggles, however, as spattered flux can be very harmful to the eyes.

Soldering methods

If you use CxM fittings to connect the small tubes to the headers, you'll need the proper size twist drill and pipe tap to make the holes and thread them. For the sizes you're likely to be using (probably ½ inch), the twist drill sizes and threading tap sizes match up as follows:

National pipe thread	Drill diameter
$\frac{3}{8}$ inch	$\frac{9}{16}$ inch
$\frac{1}{2}$ inch	$\frac{11}{16}$ inch
$\frac{5}{8}$ inch	$\frac{25}{32}$ inch

In drilling the holes for the pipe tap, it's wise to start with a "pilot hole" made with a smaller-diameter drill bit. This keeps the tip of the larger bit from wandering, and helps prevent the large bit from binding in the tube wall. Use light pressure to keep the drill speed high, and be sure all holes are aimed in the same direction.

After the holes are drilled and threaded (you can usually rent the threader from a tool-rental firm), clean the threads of the CxM fittings with a fine wire brush, brushing each one just before tightening it in place. Flux the threads you have made and also those of the CxM fitting just before turning it in place. Turn the fitting back and forth a few times to distribute the flux over the threads. Then solder it in place with 50A solder. The smaller tube is soldered into the fitting after that. Keep the assembly from moving while this is being done, as the first soldering job may be softened by the heat from the second one. Use the tools and methods described next for sweat fittings.

Soldered tubing connections

First, be sure the tubing to be connected is cut to the right length, preferably with a tubing cutter. (You can buy the cutter in most hardware stores and plumbing supply shops.) The reason the correct length is important: if a length of tubing is cut even a little too short, its end won't go all the way into the fitting, so a weak joint may result.

(In building the heat collector, correct tube lengths should not present a problem, as all tubes between headers are of the same length in most designs. Ream out the cut tubing ends.)

Cleaning is the next step. Use very fine (finishing) grade sandpaper for this, and don't overdo it. The aim is to remove only the fine film of oxide and soil from the surface. When the clean copper is visible, the job is done. Further sanding might remove enough metal to make a loose fit between tube and fitting, and interfere with the sealer to come, in this case solder. Clean the end of the tube only for slightly more than the length that will go into the fitting. And clean the inside of the fitting in the same manner. Do the cleaning job even if the tube and fitting appear clean. If you use steel wool instead of fine sandpaper, brush, wipe, or blow the cleaned surfaces afterward to clear away any steel particles that might otherwise remain and corrode.

If the paste flux is new and unopened or has been standing for a length of time, stir it up before using it, as the chemicals in it tend to settle on standing. Plan your work so the flux can be applied and the connection soldered as soon as possible after cleaning. *Do not let a fluxed connection stand for more than an hour before soldering.*

Using the torch

For the tube sizes used in the heat collector, a propane torch is usually adequate to supply the heat for soldering. If the work is being done out of doors in cold weather, however, you may need a gasoline torch or other type to supply more heat. Apply the torch flame to the fitting, but do not aim it into the capillary space between tube and fitting. The tube and fitting should be held either by clamping to a support or with pliers (which may require an assistant). Try the tip of the solder wire frequently against the capillary space where the tube fits in the fitting. When the parts have reached soldering temperature, the solder will melt on contact and flow into the joint. If the joint has been properly prepared, a glistening ring of solder will appear almost instantly all around the joint. The flux in the joint will have been displaced by the solder, but should be wiped off with a rag, as it can have a corrosive effect over a period of time. If a residue remains, you can take it off later with a rag dipped in detergent solution. The reason for trying the tip of the solder wire against the joint frequently while heating: if the joint is overheated the flux may be burned—preventing solder from entering the joint. This requires separating the joint, recleaning and refluxing. By trying the solder against the joint at short intervals during heating, you don't risk going too far beyond its melting point before completing the work.

Where the connection being soldered is close to another one already made on the same tube, lay wet rags over the nearby joint so heat traveling along the tube won't melt the solder in it.

Allow soldered joints to cool naturally for some time before applying water if additional cooling is necessary for handling. This is especially important if the fitting is a cast type, which might be cracked by too rapid cooling.

Sheet-metal soldering

If your sheet metal must be joined to make up the area required, you have a choice of various seam systems, two of which are shown in the drawings. Both are sealed by soldering after assembly. The surface of the sheets should first be cleaned, fluxed and "tinned" along an area close to the edges that will be joined. The width of this area depends on the width of the seam overlap. Tinning is simply a matter of heating the sheet close to the area that has been cleaned and fluxed, so that a coating of 50A solder can be flowed on like paint. Then, when the meeting surfaces are riveted together, or bent over and flattened, additional heat causes the tinned surfaces to soften and fuse together. Additional solder can be added during the final heating, and flowed into any gaps. The work calls for patience, but working time varies with the method. In a fair-sized example, a flat seam on a roof heat collector for a swimming pool involved 600 linear feet of seams, required 48 hours of working time and consumed approximately 60 pounds of solder. If you're unfamiliar with this type of work, it's wise to try it on a small scale before deciding to do the big job. If you're satisfied that you can do it, you can then buy a moderate amount of solder and flux and keep track of how much you use for the first few yards of seam. From that you can estimate the total amount you'll need. The amount of solder required for the tubing connections is relatively small, but can be figured from the following list:

Copper tube size (in inches) nominal diameter	Amount of solder (in pounds) required per joint
⅜	0.0025
½	0.004
¾	0.005
1	0.0075
1½	0.01
2	0.0125

Source: Copper Development Association

RIVETED AND SOLDERED LAP SEAM

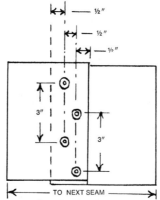

COPPER RIVETS: 5/32" DIAMETER MINIMUM (NO. 52 RIVETS)
CAN USE: 1. "BLIND" RIVETS
     2. EXPLOSIVE RIVETS
     3. SOLID RIVETS
SOLID RIVETS MUST BE USED WITH COPPER WASHERS OR "BURRS" TO
PREVENT DAMAGE TO THE PARENT METAL. CROSS SECTION OF SEAM
WOULD LOOK LIKE

SOLDER AROUND EACH RIVET
SEAM SOLDERED THOROUGHLY

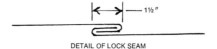

DETAIL OF LOCK SEAM

To form the bends in the locking type of seam, clamp the copper sheet (after tinning) between two straight-edged 2 × 4's or between two lengths of angle iron, with the required width of copper protruding. It is then a simple matter to bend the copper to a right angle by placing a wood block against the protruding edge and tapping it with a hammer. The copper is removed from the clamping to complete the bend to the locking form. Flux can be placed in the seams at this stage. The two bent-over edges can then be interlocked, and flattened by hammering downward on a wood block while the underside of the seam is supported on a flat, firm surface. Heating to fuse the solder follows, along with any required flow-in.

There are other effective ways of doing this part of the collector work. If you have had success with another method, it may be better for you to use it. Proper soldering of both the sheet metal and the plumbing, however, is vital to the successful operation of your heat collector. If you don't feel you can do it yourself, get an estimate from a sheet-metal shop.

Building the heat-collector case

One type of construction is shown for the enclosing case of the heat collector. (The term "case" here refers to the enclosing frame and related members, the glazing, and the insulation.) Many modifications are possible to utilize economical materials or to suit special situations, so long as they do the job. The important requirements are that the design admit maximum sunlight, withstand winter winds and weather with as little heat loss as possible, and keep the inside of the unit dry.

The methods by which the copper heat-absorber plate can be made up from available sheet sizes have already been described, along with those for tube connection. These are the methods you'll be using in the assembly of the unit. Its dimensions will depend on the factors covered in previous chapters (particularly Chapter 2) and the job you want it to do. For simplicity, a section 2 × 8 feet is used,

Ready-made Sunworks collector panels. Copper collector plates have ridges pressed into them, with tubes soldered in the ridges. Ridges are visible here through the glass cover, tubes are inside ridges, on underside of panel. Panels are available in variety of sizes.

though other section sizes can be built on the same general design to suit individual situations. The sections, of course, can be combined to provide the overall area needed. They are designed to be mounted on top of the roof decking and a roofing layer so that possible damage to the heat collector would not result in roof leaks.

If, instead of building your own collectors, you use ready-made ones, such as the Sunworks units shown, in new construction, you can select a size that fits between rafters spaced on 2-foot centers,

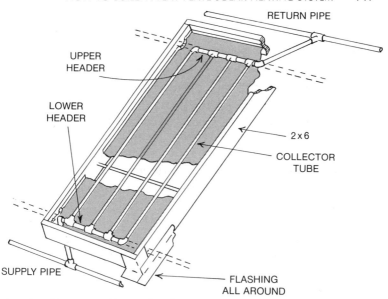

RETURN PIPE

UPPER
HEADER

LOWER
HEADER

2 x 6

COLLECTOR
TUBE

SUPPLY PIPE

FLASHING
ALL AROUND

Sample collector assembly suited to do-it-yourself construction. Where number of sections are connected as indicated by dotted lines, single 2 × 6 is shared by adjoining sections. Supply and return pipes should connect to ends of headers feeding connected sections. Heater tube arrangement shown is grid type. Continuous serpentine pattern (sinuous) is used instead in some systems, as described in text.

so that the collector plate also serves as a roof. The Revere roof-panel system described earlier also serves as both roof and heat collector, so you don't duplicate costs in a new building.

The plans shown for the complete do-it-yourself heat collector may be followed in detail, or they may serve merely as a guide. If you incorporate your own variations, keep the basics in mind, as described in Chapter 2. The collector in the drawings, including the do-it-yourself plate and tube assembly, is based on the use of 2 × 6 lumber in perimeter framing, which allows for approximately an inch of space between the tubes and the glass, and 3½ inches of fiberglass under the plate, between it and the roof surface. Intermediate cross-supports between 2 × 6's are of ¾-inch copper tube fitted in holes bored in the 2 × 6's. These prevent sagging of the plate and tubes. The cross-supports should be spaced 2 feet apart, though some variation is permissible to make spacing come out even when collector length isn't a multiple of two. Half-inch galvanized or black iron pipe may also be used if it is thoroughly coated with heat-resistant paint. In this event the intermediate supports may be spaced 3 feet apart. A single midpoint cross-support would thus be used in a collector with a length of 6 feet.

As the 2 × 6 rim framing supports the plate and tube assembly, the framing must be the first part of your heat collector to be put in place, assuming the plate and tube units are already completed. Be sure the corners are square (a large steel square is good for this) and fasten the frame with a few corner angles held temporarily to the roof with a single screw each into the roof. All screws should be driven into the 2 × 6's through the vertical legs of the angles. The intermediate 2 × 6 cross-members (top to bottom) are set in place and nailed through the perimeter frame into the cross-member ends. Use the actual width measurement of the copper plates, or pans, to mark the spacing of these intermediate cross-members.

Measure down 1½ inches at several points along the inside of each 2 × 6 and use a straightedge to draw a line at this level. Then screw a ¾" × ¾" strip along this line, as shown by the dotted line in the detail drawing on page 123. Also bore the holes for the pipe or tubing cross-supports, using a power drill and a spade bit matched to the pipe or tube diameter. Bore all the way through the intermediate 2 × 6's, but only halfway through the end ones, so there are no through openings to the outside. Also bore the two drain holes at the lower end of each section, and attach the metal baffles to the outside with short aluminum or bronze nails, as shown. These minimize snow blow-in. Make a final check for squareness, then drive the rest of the screws in through the horizontal legs of the corner angles into the roof. If any adjustments, such as correcting slight variations from squareness, are necessary, remove the temporary screws from the corner angles where necessary, make the adjustment, and then refasten with all the final screws.

The framing should be thoroughly painted at this stage, inside and out. Use a good grade of exterior gloss enamel or other weatherproof paint, and follow the manufacturer's instructions as to the priming (if needed) and the number of coats.

Holes for the tubes that lead from the plate-tube assemblies through the roof to the supply and return mains should be bored at this time. These are slightly oversize (1¼ inch for ¾-inch tube), but are rimmed with plywood rings, nailed, and roof-cemented in place. These act as dams around the holes to keep water from running into the holes if glass is broken. (The angle and position of these lead-in tubes must be carefully planned so connections can be made to the mains inside. The interior plumbing should also be accessible. More about this later.) Making the holes oversize eases installation of the plate-tube assemblies, and allows for expansion and contraction. (A 10-foot length of copper tubing expands about ⅛ inch in length with a 100-degree rise in temperature—an expansion figure that can be expected for the entire unit.) For this reason, the holes through which the fastening screws are driven through the up-bent plate sides should also be oversize. Washers under the screw heads keep the heads from passing through the holes. The screws should be snug, but not overtightened.

Before the plate-tube assemblies are set in place and fastened, the fiberglass insulation is placed over each section. A short, pointed piece of ¾-inch dowel can be inserted in the end of each lead-in tube to ease a hole through the fiberglass as the tubes pass through. This also prevents particles of the fiberglass from entering the tubes.

Once each plate-tube assembly is set in place, drive the screws that hold it, making certain that the edges are straight. If any part of an edge sags, pull it up with pliers to straighten it, and drive a brad partially through one of the nearby holes to hold it until screws can be driven in adjacent holes. Then pull out the brad and replace it with a screw. After the screws are driven, the upper edge of the copper can be sealed to the wood with a flexible sealant of silicone or Thiokol type.

Flashing

Aluminum flashing should be nailed all the way around the perimeter of the collector, working from the lower end up the sides, and ending with the flashing across the top, so that all sections overlap in shingle

OPTIMUM COLLECTOR ANGLE VERSUS LATITUDE

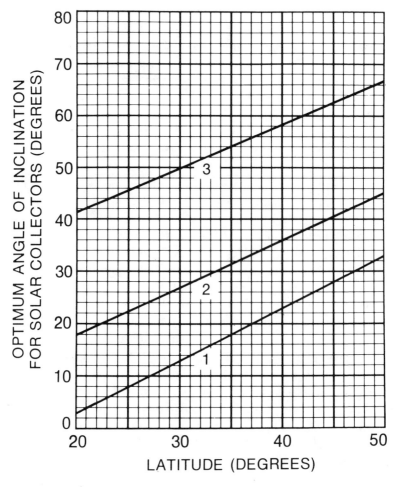

Base the inclination of your heat collector on your latitude plus number of degrees indicated on graph for winter heating. Line 1 is for summer maximum, line 2 for year-round maximum, line 3 for winter maximum, as in home heating.

fashion. It's wise to bed the flashing in roof cement, and use the same cement to cover the heads of all nails that pass through the flashing. If you work carefully you can avoid smears of roofing cement on the surrounding roof area. The flashing, however, must be used to assure a watertight joint between the collector and the roof. If the collector must be tilted to a different angle from that of the roof to bring it to the correct inclination, the flashing is placed at the juncture between the roof and the base of the angled structure.

If you buy your collectors

If you buy ready-made collectors or assemble them from ready-made components, be sure to have all dimensions and data regarding them on hand before you start work on your sun-heat system. By selecting the most suitable unit size in advance, you can often simplify your work. You can also adapt your roof plans to the available collector sizes if the installation will be part of new construction. If the collectors are to be mounted on an existing roof, and additional structure is needed to support them at the correct angle, any necessary modifications, such as bracing, can be planned and completed before the collector mounting work begins. It is also important that you know the weight of the collector units in the size you'll use, so you can plan to get them to their location without problems.

Connections to the mains

Plan the connections from the roof heat collectors to the supply and return mains inside the attic to allow adequate working space for soldering the fittings. Also make sure any valves and air vents are accessible. And allow space for insulating wrapping along all exposed sections of tubing, including the mains, to minimize heat loss in the attic. A pressure relief valve should of course be provided also. The overflow from this should be led to a suitable container open to

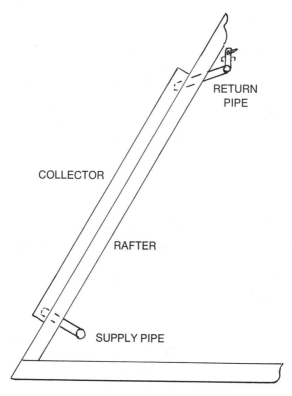

Where collectors are mounted on top of existing roof, piping is led through roof to supply and return pipes. Ready-made collectors are often available in sizes to fit between rafters so piping is inside.

the atmosphere so antifreeze will not be lost when the valve operates.

Glazing

The glazing system shown requires two widths of ⅛-inch aluminum strip. The lower one is ½ inch wide, the upper one 1¼ inches wide. A silicone sealant, available from glass suppliers, is used between the upper aluminum strip and the glass. The sealant is sold in tubes to fit a caulking gun, with which the sealant is spread in a narrow bead along the glass before the aluminum is set in place. You can set the diameter of the bead by the size of the hole you make in the end of the tube. (It comes out like toothpaste when you squeeze the trigger of the caulking gun.) A bead diameter of slightly more than ⅛ inch is usually about right. It should squeeze out very slightly beyond the edge of the aluminum when the strip is tightened down. Don't tighten the screws excessively, as unevenness in the lumber can result in cracked glass. The sealant closes minor gaps. If there's no squeeze-out at any section, work more sealant into the gap between the glass and the strip. This can be done by holding the end of the tube against the edge of the strip, and moving the gun forward as you trigger out

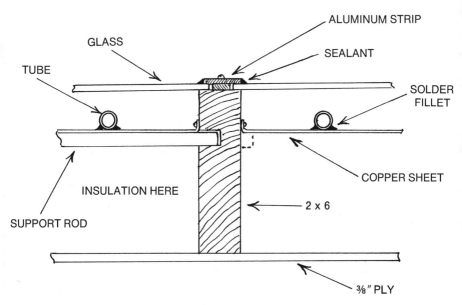

Cross-section of heat-collector assembly shown previously, at location of 2 × 6 side frame member. If units are built on roof decking, as in new construction, decking can take the place of ⅜-inch plywood shown. Insulation goes between plywood and copper sheet. Ledger strip, shown dotted, may be nailed to inside of 2 × 6 frame to ease installation of copper sheet at fixed level. Upturned edges of copper are nailed to inside of 2 × 6 frame. Design may be modified to suit situation.

the sealant. The tip of the nozzle then helps push the sealant into the gap. You can complete the job with a putty knife or flat wooden strip (like a tongue depressor). Wipe off any excess sealant that remains on the glass. The sealant is available clear and in several colors.

In most areas you won't have to wash the glass, as rain will remove dust accumulations. But check up on the glass a few months after your collector is completed. In some localities a seasonal cleaning may be necessitated by certain types of air pollution. A bucket of detergent, a long-handled sponge and a window-cleaning squeegee do the job.

The overall system

The system diagrams shown for the Revere solar-energy collectors are of types generally used in flat-plate solar-heating systems. They offer a choice of simple heating; heating plus domestic hot water; and heating, hot water, and solar-assisted air conditioning. (More about this type of air conditioning in Chapter 7.) If you prefer to start small, with domestic water heating only, two methods of doing it are also illustrated by the diagrams.

All systems are shown with a separate antifreeze circulating system and heat exchanger to transfer the heat to the water in the storage tank. The reason: in areas where winter temperatures reach the freezing point, the heat collector might freeze at night if filled with water. Although the collectors could be designed to drain back to the tank at night, this involves the risk of incomplete drainage, and possible freeze damage. In areas where winter temperatures do not drop to freezing, the systems could be modified so that storage-tank water would circulate through the heat collectors, eliminating the heat exchanger. Where heat from the water in the main heat-storage tank is used to provide domestic hot water, however, a heat exchanger must always be used in the domestic water line, so that there is no intermixing of main storage tank water with domestic hot water. Also, where winter freezing does not occur, a gravity domestic water

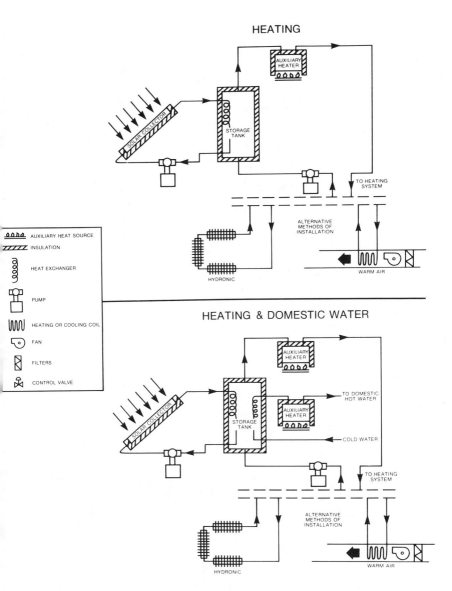

HEATING

HEATING & DOMESTIC WATER

AUXILIARY HEAT SOURCE
INSULATION
HEAT EXCHANGER
PUMP
HEATING OR COOLING COIL
FAN
FILTERS
CONTROL VALVE

AUXILIARY HEATER
STORAGE TANK
SOLAR COLLECTOR
TO HEATING SYSTEM
ALTERNATIVE METHODS OF INSTALLATION
HYDRONIC
WARM AIR

TO DOMESTIC HOT WATER
COLD WATER

Flow circuits for typical residential solar heating. Heat exchanger in flow circuit from heat collector permits use of antifreeze in heat collector. Heat from small volume of antifreeze in collector circuit is transferred through heat exchanger to 1,500 gallons of water in heat storage tank. Smaller heat exchanger for domestic hot water picks up heat from heat-storage tank and transfers it to fresh water (without intermixing) that flows to hot-water faucets.

PIPING SYSTEM DESIGN

HEATING, AIR CONDITIONING & DOMESTIC WATER

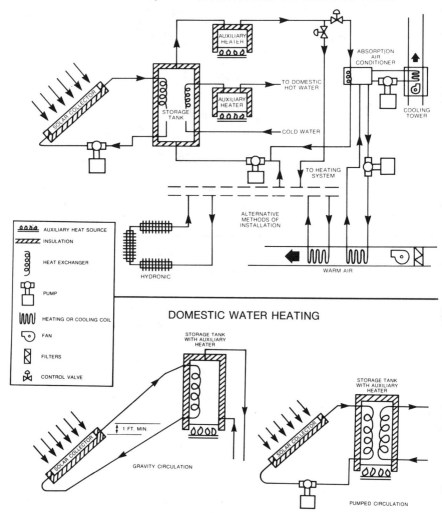

Upper circuitry incorporates air conditioning powered by solar heat. Details in Chapter 7. Lower diagrams are of solar water heaters. Type at left works by natural circulation, requires no pump, but must have bottom of hot-water storage tank at least 1 foot above top of solar-heat collector. System at right utilizes pump, can have tank below heat collector. Second heat exchanger is shown in tank, though some systems omit this and operate on same basis as heater at left. Antifreeze flows through solar-heat collector circuit in both examples to prevent winter freeze-up. In warm winter areas, domestic water can flow through heat collector, eliminating need for heat exchanger.

heater may be used without a heat exchanger, as the domestic water can be allowed to flow through the heat collector.

For the sake of clarity, some details of the systems are not included in the diagrams. These will be described shortly. In general, the circulating systems for solar heating can be compared to those of conventional hydronic (forced circulation) hot-water residential heating systems. Instead of a boiler in the basement, however, the solar-heating system has for its heat source a solar-heat collector, usually on the roof.

Like a conventional hot-water heating system, the solar-heating system has an expansion tank in the plumbing that leads to the collector. This tank, usually located in the basement, is the same type as that used in conventional hot-water heating. In its simplest form, it is a small cylindrical tank connected by a vertical pipe to a T in the main from the heat collector. A "boiler air control fitting" and "tank air control fitting" in the vertical pipe divert air from the antifreeze fluid into the expansion tank, and also provide a means of releasing excess air. An adequate volume of air is allowed to remain in the tank, however. When the antifreeze in the system expands as it is heated by the collector, the added volume caused by the expansion forces its way into the expansion tank, compressing the air above it in the tank. When the antifreeze cools and contracts at night, flowing back into the system, the air in the upper portion of the tank expands, maintaining pressure on the antifreeze. If you use a diaphragm type of expansion tank, check with the manufacturer as to whether the diaphragm material is suitable for use with the usual ethylene glycol antifreeze solution. (The chances are it will be.) The percentage of ethylene glycol in the antifreeze mix should be based on the lowest outside air temperature expected. It's very important not to exceed 60 percent antifreeze (such as Prestone), although outside temperatures would usually require considerably less. When you buy a circulating pump (as used in conventional hot-water heating systems) for your heat-collector circuit, check with the manufacturer to be sure of a model suitable for use with ethylene glycol antifreeze. (The water-antifreeze mixture is more dense than water. It also has a lower specific heat, so holds somewhat less heat than a corresponding amount of water.)

A gate or ball valve should be included in the supply-return plumbing for the heat collectors, usually just before the pump in the line of flow. This permits adjustment of the flow rate to the most effective point. In general, a flow rate not exceeding 4 feet per second is recommended. However, you're not likely to have the means of measuring the flow rate, so you'll have to judge by performance. In general, the faster your storage tank reaches heating temperature, the better your adjustment—which may call for a period of tuning up. If the flow through the heat collector is too slow, the outcoming water temperature will be high, but the volume of water returning to the storage tank will be small. If the flow is too fast, a greater volume of water will be returned to the tank, but its temperature will be relatively low during the early stages. Your tune-up task: adjust to the most effective rate. You may find your system operates well with very little adjustment. You can check the operation of individual sections of your collector by feeling the tubes (inside the attic) that lead from the collectors to the return main. If any section is noticeably cooler than the others you can suspect an air lock. So automatic or manual air vents should be included at the high points of the system. (Automatic ones save trouble, and in some cases include a manual feature.)

Connections at the collectors

Sections of the collector may be connected individually to the supply and return mains, or they may be grouped so that two or more act as a single unit. If the sections are grouped, the headers of the grouped sections are connected together so that the flow travels from the supply main through the connected supply headers (and up the tubes) of the entire group to the connected return headers of the group, then back through the return main. In average-sized collectors (like those in the photographs), individual sections or groups containing around twelve tubes are often connected to the mains as units.

Collector ratings

Although you won't have precise efficiency figures in advance on heat collectors you build yourself, you can make an educated guess by comparing them to similar units for which performance figures are available. If you use ready-made collectors or units built from kit components like the Revere heat collectors, obtain all the information available from the manufacturer. The graphs shown, for example, indicate the efficiency that can be expected of Revere collectors under a variety of conditions, for panels with either three or four tubes per 2-foot panel width, and with one or two layers of glass. The dark line on the graph for three tubes per 2-foot-wide panel with a single layer of glass is an example that demonstrates the ability of this type of collector to furnish heat for industrial applications far beyond the requirements of home heating or pool heating.

A total system capacity of 250,000 Btu per hour is required from 2,000 square feet of collector area, with an average heat-collector-panel water temperature of 120 degrees F, a solar heat input of 250 Btu per hour per square foot, and an ambient temperature of 80 degrees F surrounding the collector.

If you divide the required 250,000 Btu by the 2,000 square feet of collector area, you find that you require 125 Btu per hour per square foot. So you need to capture half of the 250 Btu per square foot that is your input, or 50 percent efficiency. The graph shows (as indicated by the dark lines) that a 2-foot-wide panel with three tubes and a single layer of glass has an efficiency of 52 percent and an output of about 130 Btu per square foot, so it can do the job.

As a number of variables affect the panel water temperature, you can use the graphs to show the efficiency you can expect based on several assumed water temperatures, starting with an estimated one, say, of 100 degrees. Then refigure with a higher one like 125, as the collectors can normally produce water temperatures 100 degrees above the surrounding air temperature. If a solar-heat collector system is operating in your area, its panel water temperature will be a guide. As pointed out in Chapter 2, even advanced mathematics

SOLAR ENERGY COLLECTOR SYSTEM RATINGS
3 TUBES PER TWO FEET WIDE PANEL

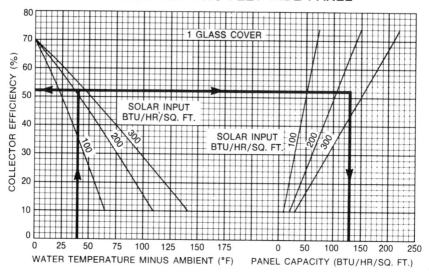

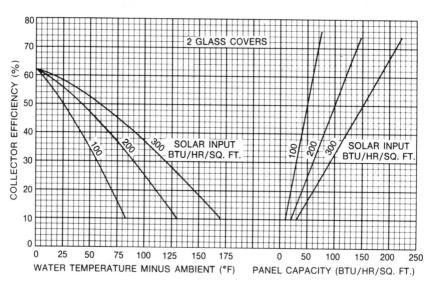

To figure collector efficiency of Revere rectangular-tube solar collectors, use these graphs. Note that the efficiency increases with the number of tubes per panel. Single layer of glass is usual. Extra layer is used for severe winter conditions. Dark line in upper left graph is explained by example in text.

SOLAR ENERGY COLLECTOR SYSTEM RATINGS
4 TUBES PER TWO FEET WIDE PANEL

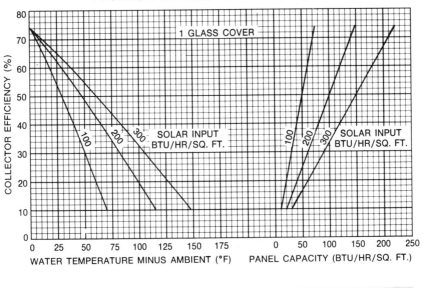

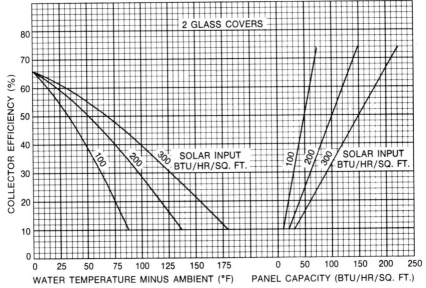

NB THE ABOVE FIGURES APPLY TO THE HEATING OF WATER.

can't tie down unpredictable factors like weather and wind velocity. So your final performance figures will come from actual testing. If you are using ready-made collectors of another make, of course, base your estimating on the specifications of the manufacturer.

If you start small, as with a sun-fired domestic water heater with around 50 square feet of collector area and a storage tank of around 80 gallons capacity, you can do some testing before you tackle a larger installation. But comparisons between a small and a large system should take into consideration factors that vary with size. The small tank, for example, is likely to be completely heated in a relatively short period. But as the temperature rises during the heating process, the difference between the tank temperature and the temperature of the water from the collector is narrowed, and heating is slowed. But in general, if your small-scale installation shows good performance, solar heating is likely to be feasible. If you have the equipment to take accurate measurements of the temperature rise at timed intervals, you can establish a basis for planning a larger system. If your water-heating unit is equipped with a circulator pump and valves, you can also experiment with flow rates.

The backup system

If your solar heating is installed in an existing house, the conventional heating system already in it can usually serve as the backup. In this event, the solar heating operates as an independent system, though the automatic controls of the two systems may be linked. (More about this shortly.) In some cases, too, part of the existing heat-distribution system may be utilized by the solar heating to avoid duplication and to reduce overall costs of the solar heating. If, for example, the present heating system is of the warm-air type, and if the ducts are sized to accommodate central air conditioning, you can use the existing ducts for your solar heating's distribution system. If the ducts are sized for warm-air heating only, you'll probably need to replace them with larger ones.

If the existing heating is of the hot-water–baseboard-radiation type, the radiation area would usually have to be increased considerably for use with the lower-temperature water from the solar-heating system in order to do the same heating job. As this is frequently both difficult and uneconomical, an independent duct distribution system with a fan-coil unit can be used with the solar heating. The existing hot-water system then functions as the backup.

Controls

The controls that automate the overall heating combination (solar and backup) are readily available, though some may have to be ordered, as they may not be stocked by local suppliers. The type of thermostat that has proved best suited to the circulating system between the heat collector and storage tank, for example, is usually called a differential thermostat, and is not required by conventional heating systems. Unlike the familiar wall thermostat, it has two temperature-sensing elements that can be located at a distance from each other. One of these is placed in the heat collector, the other in the heat-storage tank, near the top. Unless the temperature in the heat collector is higher (typically about 5 degrees) than the temperature in the tank, the thermostat will not switch on the pump that circulates fluid (such as antifreeze) through the heat-collector piping. This assures that the pump cannot operate when the heat collector (being colder than the tank) would actually cool the water in the storage tank instead of heating it. Thus, this thermostat shuts off the pump when the heat collector cools down at the end of the day and during cloudy weather. It also shuts off the collector pump under certain conditions of backup operation, as described later. A typical thermostat of this type is the Honeywell L-643A. Check with the manufacturer on installation and adjustment details.

The wall thermostats that actuate the solar-heat distribution system to send heat through the house and the backup system that takes over when the solar heating can't maintain the set temperature vary with the design of the systems. The general principle: the backup system's thermostat is set at a slightly lower temperature than that of the solar heating's thermostat. (A single thermostat with dual temperature settings for this type of operation may be used instead of two separate ones. A typical thermostat for this purpose is the Honeywell T872. Check with the manufacturer for details.) Thus, if the room temperature continues to fall with the solar heating in operation, it will eventually drop to the switch-on setting of the backup thermostat, and the backup system will go into action. If the two heating systems are integrated (the simple heating diagram is a good example), hot water will then flow from the backup heater through the heat-distribution system (warming the house) and on into the storage tank (increasing tank temperature), and finally back through the backup heater.

While this is taking place, the separate heat-collector circulating system pump will continue to operate so long as the temperature in the collector is higher than that in the tank. When this condition is reversed, the differential thermostat automatically shuts off the collector circulating pump. When the inside temperature of the house reaches the upper level set for the combined systems, the backup system's thermostat shuts off the backup heater. When the temperature again falls to the switch-on temperature of the solar heating, the system's heat distribution starts. If the temperature set on its thermostat is maintained, solar heating provides the heat without the backup system. If not, the backup system cuts in, as just described, and the cycle is repeated. This interoperation of the two systems is sometimes called the "modulating" method.

In other control arrangements, especially when the solar and backup systems are not directly integrated, another method is often used. This employs a thermostat that automatically shuts off the heat-distribution system of the solar heating simultaneously with switching on the backup system. (Because of the double action, this is sometimes called the "bang-bang" method.) The backup system, being independent of the solar heating, then takes over the house heating completely, allowing the solar collector to build up heat in the storage tank during the period of backup operation. This method can be applied to a variety of solar-backup combinations. An example: where an existing hot-water baseboard system serves as the backup for a fan-coil solar-heating system. It can also be used in other combinations where the flow of hot water from a backup heater to the heat distribution system (whatever type) can be separated from the storage-tank system, as by electrically operated valves. Since a smaller volume of water is heated and circulated by this method, a more rapid room-temperature response is possible.

System flow details

In planning your piping, make it as simple and direct as possible. Keep the individual circuits at reasonable lengths to prevent excessive pressure drops or high temperature rises in each of the circuits. And provide air vents at the high points of the system to prevent air locks. Avoid "short circuiting," such as might result if the circuit through one portion of the heat-collector system was much shorter than another. And include "balancing" valves to permit adjusting the flow through the individual circuits (V gate valves or globe valves).

If you don't have a use for the entire heat-collecting system in the summer (as for pool heating), plan the piping in separate sections so that some of the heat collectors can be cut off by valves and drained through others. The antifreeze solution can be saved and reused when the cut-off portion of the system is refilled and reactivated in the fall. The part of the solar-heating system that remains in operation through the summer months can be used to provide domestic hot water. If a heat collector remains filled in summer, but is cut off from the circulating pump, it is likely to reach boiling temperatures and lose fluid through the pressure relief valve. (As outlined in Chapter 7, solar-powered air conditioning can utilize the heat from collectors in summer if suitable air conditioning equipment is available.)

Filling the system

The water in the heat-storage tank should be as free as possible of elements likely to cause corrosion or scale inside the system. If the local tap water causes no problems in conventional hot-water heating systems, it should be suitable. If you have doubts, have it tested before using it. Many storage tanks (time and weather permitting) are filled with filtered rain water, which is actually distilled water. One inch of rain on 1,000 square feet of roof gives you about 623 gallons. So a few heavy spring rains can usually fill your storage tank if you channel the water in from the rain gutters.

There are several methods for filling the heat-collector circuit with antifreeze solution. Even though the circulating pump may have ample power to circulate the fluid through the system, once the system is filled, it may not have power enough to drive fluid to the top of the system during the filling process. In this event, a large centrifugal pump may be used to fill the system, drawing the anti-freeze-water solution from a barrel and forcing it into a valved T. When the fluid comes out through a second valved T and flows into the barrel, the valves can be closed gradually, and the system is ready for operation. The regular circulating pump can take over after the system is filled and the hoses are disconnected. Another method sometimes used calls for using the large centrifugal pump to fill the system through a valved T at the top of the system, allowing air to escape through a second valved T, also at the top of the system, until fluid comes out of the second valved T. The valves are then closed, as mentioned, and the system is ready for operation with the circulating pump.

Other system types

If you build a system other than the flat-plate collector type, or if you combine several types (as flat-plate and south window wall), keep an accurate record of performance if you can. It will be useful in tuning your solar heat for maximum performance. And it may help a future builder.

Whatever system or combination you use, take advantage of any improvements possible as they come along. As shown in the drawing, for example, it's a simple matter to make insulating panels from

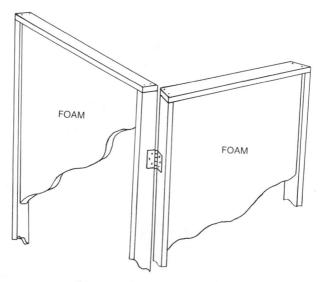

FOAM

FOAM

Foam insulating panels for large window areas may be made from nominal 1 × 2 lumber and expanded polystyrene foam in either 1-inch or 2-inch thickness. Bend foam to frame only with a glue suited to foam, such as Dexcel 161-D, made by Dacar Chemical Products Co., Pittsburgh, Pa. 15220.

Sunworks solar-heated house, with heating system designed by Everett M. Barber, Jr., is located on New England shore. One bank of heat collectors can be seen on top of the house to the left. Large window area facing south contributes further to the solar heating. Photo, taken on early-spring morning, shows how roof and balcony overhangs will shade windows in summer. Architect: Donald Watson.

How three banks of collectors are used on housetop for a total area of 400 square feet, to provide approximately half of home's winter heating. During starting months of first season's operation, system provided 100 percent of heating.

readily available foam, to cover large window areas at night or during cloudy periods. These enable you to utilize sun heat when it's available, while minimizing heat loss through the window area at other times. And, of course, the insulating panels can be planned so that a portion of the window area can be left exposed for light.

General tips

If there are any flat-plate heat-collector solar-heating systems within reasonable distance from your location, data on their specifications

View of two front banks of collectors of the house shown. Note that second bank is on higher level than first, so one doesn't shade the other. Rear bank, not visible, is set to be clear of second bank's shadow.

Inside house, underside of heat collectors is insulated by fiberglass with polyethylene sheet under it to act as vapor barrier. Look closely at lower ends of rafters and you can see short vertical tubes that connect lower end of each collector section to black (insulated) lower main of system.

and performance, if obtainable, will be very helpful in planning your own system. (As more solar-heating systems are built, information about them and their performance in specific areas and climates will become available as a guide to other builders. This type of information is scant at present.) Otherwise, use as a general guide the information on the houses shown. The charts can also serve this purpose, as can the information in other chapters. Remember that you are working in a field where new developments are frequent. One of them may be yours.

Modes of operation

Keep in mind that variations in the plumbing of your solar heating can be planned to utilize the heat from the collector in several ways. For

Heat-storage tank in basement of Barber Sunworks house. It holds 1,500 gallons of water, is 10 feet long, 5 feet in diameter. When work is completed it will be completely insulated by vermiculite.

With manhole cover removed from tank end, heat-exchanger coil for domestic hot water is visible. Much larger heat exchanger in bottom of tank transfers heat from antifreeze in collector system to water in tank without intermixing.

example, if the day is sunny and the house does not require heat from the collector system, as might be the case if a large south window area was part of the overall system, all heat from the collector may be sent directly to the storage tank. Conversely, if the day is sunny and the house needs all the heat possible, as on an exceptionally cold day, all heat from the collector may be sent directly to the house heating system. In test work you can do this with manual valves. In a permanent installation, thermostatically operated valves may be used. Where a fuel-fired water heater or boiler is used as a backup, it is also possible to provide a by-pass pipe and valves to permit heat from the backup to be sent directly through the house heating system without contributing heat to the storage tank. This might be advantageous under some extremely cold weather conditions when quick

Fan-coil unit of Barber-designed solar-heating system rests temporarily on cement blocks. Air is drawn in by blower through finned radiator in foreground. Hot water from heat-storage tank, flowing through radiator, heats air as it passes through. Air continues on through ducts (as at top of photo) to outlets in rooms of house, as in conventional warm-air heating system.

All ducts, including branch ducts, are larger than those in conventional warm-air heating system. These are about 20 percent larger.

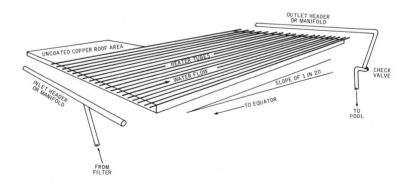

Flow diagram of swimming-pool heater mentioned in text. Size of collector depends on size of pool. Can be used without glass cover in warm-climate areas.

Swimming-pool heater built on garage roof.

house heat is needed. Although such arrangements may not be required, they're worth thinking about to meet special situations.

If you build an experimental collector

If you'd like to gather some firsthand information on the actual operation of a heat collector before you tackle the construction of a full-sized sun-heating system, you can build a relatively small experimental unit. If you do, it's wise to make it of a size that can serve some useful purpose, such as water heating, after the test period. It might also be designed to become part of a larger system if you decide to follow up with one. Your experimental unit, however, should be planned to facilitate the type of modifications you may want to try. For example, you might want to start with a minimum number of tubes between headers, and later increase the number to check on the resulting increase in output. Or you might try one, then two layers of glass, a layer of plastic under a layer of glass, perhaps plastic alone. All these variations were tried in the experimental collector shown in the photos. From this type of testing you can often determine the least expensive form of collector likely to meet your particular needs.

To keep costs down, shop for your materials. Some of those used in the unit shown were bought secondhand from building wreckers and similar sources, others were bought new because they were unavailable otherwise. In some cases you may be able to buy tubing

Ladder makes a good base for outdoor construction of test heat collector. Here, supporting strip for copper backing plate is being nailed to side piece. Be sure side and end pieces are straight lumber. Board on right side of ladder shows crook to pronounced degree, is suitable only for short parts.

Assembling collector frame. This one is experimental, built from nominal 1 × 6 lumber. Hardboard panel at right will be used for squaring frame, also for bending up edges of copper back plate. Hardboard is cut $^3/16$ inch smaller than inside of wood frame, so copper can be bent over its edges for close fit.

Soldering tubes to sheet-copper backing plate. Do it on top of strip of asbestos cement board to protect grass (or floor) and prevent rapid dissipation of torch heat. In cold weather do this job indoors, preferably on cement basement or garage floor. Follow instructions in text as to solder, flux, torches, etc.

After tubes and sheet strips of plate are soldered (and tubes connected to headers), paint the entire metal area with flat black paint. As this collector is for testing, wood frame is left unpainted at this stage.

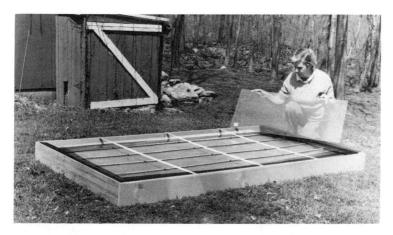

For testing in moderate weather, collector may be covered with either glass or plastic sheet. Plastic is cheaper but not as effective in trapping heat. If you buy glass from building wrecker for economy, space your supporting framework to fit the sizes you get.

cheap if you're willing to use cut pieces of various lengths. Such pieces, even for the collector tubes, can be utilized simply by connecting them with standard couplings. The coupled sections can be soldered to the backing· plate almost as easily as continuous lengths, as the gaps between tube and plate at the ends of the couplings are small enough to fill with solder, with a little patience. In the collector shown, the original tubes were full length. Later, when the number of tubes was increased by adding additional ones between the originals, most of the added tubes were made up of coupled pieces.

To add the new tubes, sections of header were cut out somewhat longer than the length of the cross-member of the T fittings needed to connect the new tubes. The T in each case was then connected and soldered to the small tube and to the header at one side of the cross-member. The gap at the other side was filled in with a short section of header-size tube inserted into the T and soldered in the usual manner after cleaning and fluxing. The other end of this small section of tube was then connected to the header proper, by means of a "slip coupling." This is a coupling that has no internal ridge to prevent it from sliding beyond the normal connection depth. Essentially, it's just a piece of tube sized to slide over the tube being connected. Thus, it can be slid onto the header tube far enough to be completely out of the way when the parts are assembled. Then it can be slid into place over the juncture of the existing header and the short section of header-size tube, and soldered. To make this easy in the experimental unit, the last few inches of the original tubes were left unsoldered (to the plate) to permit some flexing when the new tubes were added. Later, these portions were soldered to the plate. Cloth-backed abrasive was slipped between tube and plate, and pulled back and forth to remove paint and oxide before fluxing and soldering. Paint was also removed from the plate, of course, to permit cleaning and fluxing along the lines where the new tubes were added. The usual soldering methods were used. While this type of experimental procedure requires some time and work, it can often save considerable money and effort in the long run, as it gives you a clear idea of the performance you can expect in your particular area. And it provides experience in heat-collector construction.

You can try it on the ground

As shown, the experimental unit was first tried on the ground, then on the roof. The tank was a 42-gallon one of the type often used with residential well-water systems. For this kind of test work you can usually find such a tank at a bargain price at a building wrecker's yard. If you plan to use the unit for domestic water heating, however, it's better to buy a new tank.

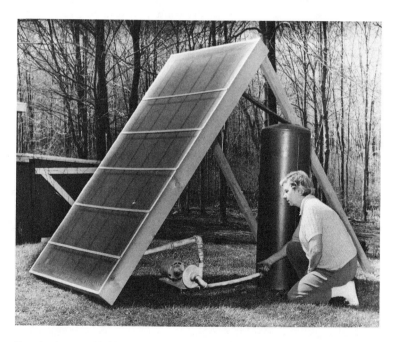

Completely assembled heat collector here is being given on-ground tryout with centrifugal pump and 42-gallon tank. Valve at base of tank is used to adjust rate of flow. If you test experimental collector on ground, you can simplify piping connections by using auto-radiator hose and clamps, also auto-heater hose.

You can adjust the collector to the correct angle of inclination on the ground merely by shifting the supporting struts nearer or farther from the collector base. An ordinary carpenter's level and a large protractor will gauge the angle. If you are likely to do extensive work with your experimental unit, however, especially if it's likely to be moved frequently, there's a handy tool for setting the angles to the horizontal, and many hardware stores stock it. The tool consists of a small level mounted on a base calibrated in degrees. Placed against your collector at the desired angle setting, it lets you see just how much adjustment is required (flatter or steeper) before you lock the collector in position, as on a roof. This type of tool was used to set and check the inclination of the collector shown, both on the ground and on the roof.

To reduce working time in the experimental plumbing, auto radiator hose was used in many of the connections between the pump, tank and collector. The hose easily withstands the temperatures likely in the unit, and is easy to connect and disconnect, using standard auto radiator hose clamps. It also permits considerable variation in the angles at the pipe connections made with it.

You can use an ordinary kitchen oven thermometer from the hardware or dime store to measure the air temperature inside the collector case. Just place it face up so you can read it through the glass. For water temperatures at the important points in the system, you can buy thermometers from plumbing suppliers. Or you can use auto cooling-system gauges bought from an auto wrecker. These aren't as widely available as they once were, and they may not be as accurate as the plumbing types, but they're likely to be less expensive. If you use the plumbing type, however, you need merely screw it into a standard threaded T fitting anywhere you want it in your piping system. One at the inlet and one at the outlet of the collector will give you a good idea of the temperature changes with different rates of flow, and at different periods of the tank warm-up. The flow-adjusting valve in the unit shown was in the main line next to the base of the tank, just before the pump.

If you want to measure the time it takes for your collector to bring the entire tank (of the approximate size shown) to a given tempera-

Sliding collector up ladder used as a ramp. Bottom of collector has been covered with hardboard to protect fiberglass insulation from the weather.

ture, of course, the tank should be insulated as it would be in normal use.

If you build your collector out of doors, like the one in the photos, you can use the same general methods. For the soldering, however, it's wise to plan outdoor work for a reasonably warm part of the year. In cold weather, soldering of this type is not easy, as the large metal areas of the plate conduct the torch heat away rapidly, and may make it necessary to use two torches.

To test the heating possibilities of a small pump-operated sun system, you can use a larger tank, say around 80 gallons (also a size often available from house wreckers), and leave it uninsulated. The heat radiated from the tank may be enough to keep a small, well-insulated room comfortable during the day and the early portion of the evening if the collector area is adequate. (Hot-water tanks con-

On-roof tryout. All piping is enclosed in collector. For permanent installation, ends of collector area can be covered.

nected to coal and wood ranges kept many a kitchen cozy in times past.) If the on-ground performance of your experimental system indicates possibilities in this direction, you can try it, preferably in chilly but not freezing weather. A garage or enclosed breezeway might be the test area.

A final thought: keep in mind that experimental sun-heating units are not limited to the flat-plate category. Many other forms of sun heating described in Chapter 3 lend themselves to small test forms that might later be incorporated into a larger system. And some, like those using air instead of fluid for heat transport, may be considerably lower in cost. Just pick a type that's promising for actual use in cutting your fuel costs. And remember that a combination of types may work together.

7 Things to Come

The things that can eventually make our houses completely solar-powered are, for the most part, already here in test form. Solar heating, of course, is well established, and growing in terms of numbers. Electric power produced directly from sunlight by cadmium sulphate (CdS) solar cells is being used for household purposes in experimental solar homes like the University of Delaware's Solar ONE house. The generating system is entirely practical from the technical standpoint, but because of the high cost of the solar cells, it's not yet suitable for general use. It is more promising economically at present, however, than the more efficient silicon cell. In bright sunlight, a typical 3-inch-square cell of the type used in Solar ONE can generate about $^3/_{10}$ watt on a steady basis. By connecting groups of cells in series or parallel, almost any variety of power values can be obtained. Current from the cells is used to charge batteries that provide a direct-current supply day and night for lights and a variety of appliances.

For appliances that require alternating current, like the average refrigerator, the current is converted by equipment built into the system. The CdS cells that produce the current are constructed so that sun heat can also be collected from their inner surfaces to augment that supplied by flat-plate collectors. (If you want to experiment with solar-generated electricity on your own, silicon-type solar cells in panels capable of supplying about 30 watt-hours a week are currently in experimental sizes at scientific supply houses like the Edmund Scientific Co. Other equipment, such as small motors that

Honeywell's solar-heating laboratory built to travel throughout the country for on-the-spot solar-heating tests. As large as a small home, unit is heated by solar-energy collector shown, tilted to correct angle for location of testing operation.

can be operated by the cells, is also available. Prices should be checked in advance with the supplier.)

Special salts for better heat storage

By using "salt hydrates" and related materials that melt and freeze at a much higher temperature than water, a far greater amount of heat can sometimes be stored in a given space than is possible with water. This fact is the basis for experiments now under way in heat storage. The explanation lies in the fact that much more heat is involved in changing a substance from one state to another, as from water to ice, than in merely heating or cooling it without changing its state. You need only 1 Btu, for example, to raise the temperature of 1 pound of water 1 degree Fahrenheit. But you need more than 140 Btu to change 1 pound of ice to 1 pound of water at the *same* temperature—without changing the 32-degree temperature at all. That's why, in melting, ice is about 140 times as effective in cooling a refrigerator as the same weight of water at the same temperature.

And there's another side to the story. When you freeze water to make ice, it *gives off* the same amount of heat that it *takes on* when ice melts. The idea of water giving off heat as it freezes may seem strange, but plant growers have known it for a long time. They utilize the phenomenon by placing a number of pails of water in vegetable cellars on cold nights to keep the temperature from dropping too low. As the water in the pails freezes, it gives off enough heat to protect the plants. The heat involved in the process is called the *heat of fusion.* By selecting a chemical or salt combination that melts and freezes at a temperature suited to solar heating, experimenters are taking advantage of the extra heat storage possible by utilizing the heat of fusion. The same procedure is also being tested for storing "cold" in relation to solar air conditioning. For the most part, the salts and chemicals involved are inexpensive. The problems being solved are of a technical rather than an economical nature, such as corrosion and the behavior of the chemicals during changes from one state to another.

Solar air conditioning

Using the heat of the sun to operate refrigeration equipment is not new, but novel methods and systems are opening up possibilities. The basic idea was demonstrated almost a century ago in France, when steam from a sun-heated boiler was used to operate an early version of the absorption-type refrigerator, which actually produced a small amount of ice. Later, another solar-powered steam system

SOLAR HEATING/AIR CONDITIONING LABORATORY

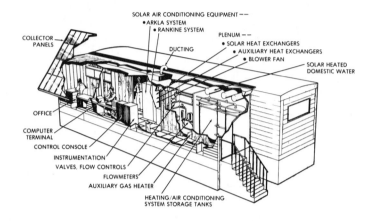

Interior of traveling solar heating–air conditioning lab.

turned out more than 500 pounds of ice per day. But the paraboloid reflector that concentrated the sunlight on the boiler, and the sun-following machinery required to operate it, were much too expensive and complex for general use. And the concentrating reflector could not produce steam in hazy weather, as diffuse sunlight can't be focused. Using a flat-plate collector system, however, the University of Florida's Solar Energy Laboratory, under the direction of Dr. Erich Farber, has demonstrated in recent years that concentrating collectors are not essential to sun-powered refrigeration. Flat-plate collectors as small as 4 feet square have produced as much as 80 pounds of ice per day with absorption refrigeration units. Larger-area systems for air conditioning have shown capacities as great as 5 tons. This is the cooling equivalent of melting 5 tons of ice per hour, with each ton removing about 12,000 Btu of heat from the room air during that hour. Sun-powered absorption air conditioners are also being used to cool the Copper Development Association's Decade 80 house in Tucson. So it's reasonable to assume that absorption air conditioners will become available for residential use in the near future, and knowing how they work will help you in planning their installation and use.

Cooling by boiling—how air conditioners work

The principle by which heat can produce cold is easy to understand if you are prepared to do a little reverse thinking. For all common types of air conditioners do their cooling by boiling a refrigerant. So to begin, think of a pot of water being heated on the stove. If the burner is turned all the way up, the water heats rapidly until it reaches 212 degrees F and starts to boil. Once it's boiling, a kitchen thermometer will show you the temperature remains at 212 degrees no matter how much heat you put into the water. Obviously, all that heat has to be going somewhere, and it is. It is going into the steam rising from the pot. By boiling the water you are changing its state from a liquid to a gas (steam) and the steam contains the extra heat, called the *heat of vaporization.* And here your reverse thinking really begins.

Instead of the burner putting heat into the water, let's say the water is stealing heat from the burner, and that heat is being carried away by the gas (steam). In a sense, you are cooling the burner. While this may not seem like refrigeration, let's suppose the pot contains a liquid that boils at a much lower temperature than water, say at a temperature *below zero.* Then it will boil and change from a liquid to a gas even at room temperature. The warm air around the pot of liquid then takes the place of the burner, and the boiling liquid steals heat from the air—heat that is carried away by the gas that results from the boiling. Now we have refrigeration in the true sense. And there are many liquids that boil at such low temperatures. We call them refrigerants, and ammonia was one of the earliest, with a boiling point approximately 33 degrees *below zero centigrade.* When it boils it steals heat from its surroundings effectively enough to turn water to ice.

In refrigeration equipment, of course, the gas that results from the

Solar-heated and air-conditioned skyscraper of the 1980's, in artist's rendering. One-and-a-half-story test model now being erected at PPG's research center at Harmar-ville, near Pittsburgh. Project will follow with larger test buildings, finally high-rise office buildings. Initial financing is by Oliver Tyrone Corp., PPG Industries, Standard Oil Co. (Sohio), and Aluminum Company of America.

boiling refrigerant can't be allowed to escape to the air for practical reasons and because it's too expensive to waste. (The ammonia, for example, is true liquid ammonia, not the highly diluted household ammonia, which is largely water.) The easiest way to understand how the refrigerant is used is by following it through a refrigeration system. And the system that illustrates it most clearly is the very common compression refrigeration system found in most home refrigerators and air conditioners. (This is not the same as the absorption type used in sun-powered air conditioners, but it makes them more readily understandable when they are described later. And both do their cooling job by boiling a refrigerant and keeping the resulting gas contained.)

The refrigerant does its boiling inside the "cooling coil" of the air conditioner, which is located inside the house. This is the coil past which *room* air is driven by a blower. As the air passes over the coil, the refrigerant that's boiling inside the coil (being much colder than the room air) soaks up heat from the air, and you feel a chilled breeze coming from your air conditioner. After taking the heat from the air of the room, the gas from the refrigerant is sucked out of the cooling coil by the compressor pump, and pushed into the "condenser" coil located outside the house. The heat soaked up inside the house goes into the condenser, carried by the gas. The condenser then acts as a radiator to dissipate the heat to the outside air. But there's more to the story.

In order for the gas to give up its heat to the outside air, it must be hotter than the outside air, as heat always travels from a hot area to a cooler one. And although the gas comes from a boiling refrigerant, the boiling takes place at a very low temperature, and the gas that results is *not* hotter than the outside air. So it has to be compressed in the condenser to concentrate the heat that's in it. The compresser does this without adding any heat. It simply crams the existing heat into a smaller space, which intensifies it. Roughly speaking, if you compress a gas to half its original volume you double its temperature, because the original amount of heat is jammed into half as much space. (If you've ever pumped up a bike tire with a foot pump you may have noticed the effect of compression. The base of the foot pump becomes very warm from the compression of the air during the pumping.) So, by compression, the refrigerant gas becomes hotter than the outside air and gives up its heat through the walls of the condenser to that outside air. In order to compress the gas, however, the compresser must have a resistance to work against. Otherwise the gas would simply flow freely through the system without being compressed. The resistance is provided by an extremely small-diameter tube called the capillary in the return line from the condenser to the cooling coil. When the gaseous refrigerant is both compressed and cooled in the condenser, against the resistance of the capillary, it returns to its liquid state, and literally squirts through the capillary into the cooling coil. Here, it boils again, and continues the cooling cycle. In operation, the entire process takes place on a continuous basis. If you shut off an air conditioner or a refrigerator that has been running for a while, and listen carefully, you can usually hear the gurgling sound made by the boiling refrigerant. If you think of the gas that boils off of the refrigerant as simply evaporating, you can also think of some everyday examples of the effect. When perspiration evaporates on your skin you can feel its cooling effect. If you dab your skin with alcohol you feel the same kind of cooling as it evaporates.

The flame that freezes

In an absorption-type unit, heat, instead of a compresser, provides the driving force that moves the refrigerant through the system. In many conventional models the heat is provided by either a gas flame or an electric resistance heating element. The refrigerant "absorbed" in water is heated in an enclosed metal chamber called a generator, and bubbled up a tube in much the same way that coffee is bubbled up the tube in a percolator. The tube leads to a chamber where the water separates out and drains back into the system, while the refrigerant continues on through a heat-dissipating section to the cooling coil, where it boils, just as in the compresser system. Similarly, it soaks up heat from the air around the cooling coil, and then passes

Decade 80 solar house being built in Tucson, Arizona, area by Copper Development Association, Inc. It will use solar energy to supply approximately 60 percent of its total energy consumption. Rooftop collectors are of the Revere type shown in Chapter 6.

on to a stage where that heat is dissipated through a radiating unit to the outside air.

There are, of course, many variations in the design and chemistry of absorption systems, but the general principle is the same. In early refrigerators, ammonia absorbed in water was often used as the refrigerant, aided in certain reactions by other elements. The solar-powered units in the Decade 80 house, however, are of the lithium bromide–water type, modified for hot-water "firing." Water heated to 225 degrees F in flat-plate sun-heat collectors provides the full-rated cooling capacity, although the units can operate with a "firing" water temperature as low as 190 degrees F. The chemistry of the refrigerant makes the low-temperature operation possible. (The manufacturer of the test units is Arkla Industries, Inc., Arkla Plaza, 400 East Capital, Little Rock, Arkansas 72203.)

Two units are used in the Decade 80 house, with a total cooling

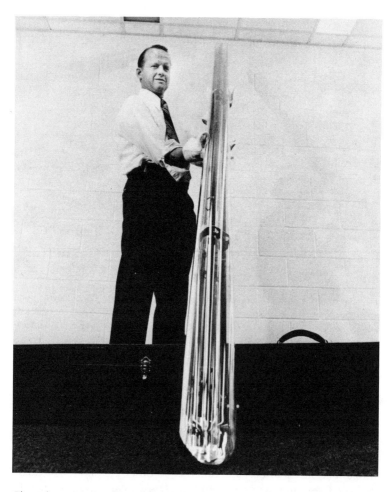

Glass tube-type heat collector being developed by Corning Glass Works. Two concentric tubes with vacuum between insulate fluid-conducting pipe inside center tube. Vacuum insulation produces high efficiency. Banks of tubes will be used in solar heating.

capacity of 72,000 Btu per hour, the equivalent of melting 6 tons of ice in the same period of time. The sun-heated water for the air conditioners is stored in a 3,000-gallon tank. To boost the temperature of the water going to the air conditioners (when necessary), a standard commercial auxiliary water heater is provided, with a capacity of up to 170,000 Btu. The heat storage and backup provisions can, of course, be used for winter heating as well as summer cooling. The collectors are of the rectangular-tube type shown and described in Chapter 6, and glazed with Pittsburgh Plate Glass "Twindow" double-glass panels.

The solar-collector area is about 1,800 square feet, based on cooling as the controlling factor. The collector panels used to warm the swimming pool will not be covered with glass, as they are to be used in the warmer months only. Ethylene glycol–water solution is used to carry the heat from the collectors. This is circulated at the rate of .45 to .75 gallon per hour per square foot of collector area.

Although the house contains 3,200 square feet of living space, making it considerably larger than average, it can serve as an example of solar heating and cooling possibilities in the same general geographical area. To prevent overpressure from excessive temperatures in the storage tank or backup heater systems, a pressure-relief valve is provided, set at 10 psig. (The g indicates "gauge," meaning the pressure as shown on a gauge would be 10 pounds per square inch above atmospheric pressure.)

8 Decorating Tips for Solar Houses

Of all the solar-heat types, the south-window-wall version is probably most likely to have a direct effect on your interior décor. But problems don't arise if you think in practical heating as well as artistic terms and strike a compromise. Usually, you'll be dealing with large double-glazed picture windows or floor-to-ceiling insulating glass that faces south or nearly south. Here, even if the original intent was purely decorative, making use of the solar factors can now put money in your bank account. Blocking a flood of bright sun on cold winter days by drawing the drapes just doesn't make sense at today's fuel prices. Letting it in and harnessing it is the modern answer. Properly done, the scheme that works best for your sun heating also tends to eliminate glare even on the brightest days. And it makes south-facing rooms especially attractive.

Although sunlight streaming in through a large south window area contributes an appreciable amount of heat to your house in any event, you can make the most of it with colors that absorb the heat and radiate it into the room. Think of rich, deep colors for your carpeting and upholstery. If you like bare floors, consider dark-toned tiles like the deep red of quarry tile, or perhaps Vermont slate. Other possibilities are brick or even dark polished hardwood. Concrete-slab floors can be painted in comparable tones. Once heated, slab floors (if that happens to be the kind you have) retain their heat for quite a length of time, gradually dissipating it into the room, and helping to keep you comfortably warm. Like the concrete sun-heated wall of the Trombe-Michel system described in Chapter 3, a slab floor can serve as a medium of heat storage. If you happen to have it, use it.

Dark paneling on walls bathed in the light of the sun can also make a major contribution to your daytime heating. Just place a small sample of dark paneling in the light of the sun coming in through any south window, and you can feel the warmth in minutes by touching it. That kind of heat spread over an entire wall can have a major effect on the temperature of a room. And the paneling creates a smart, luxurious look in living rooms, dining areas, dens, and libraries. In other rooms, such as bedrooms, where wood paneling might not be appropriate, paint, wallpaper and paneling materials suited to the situation are attractive and easy to care for. You need not carry the dark tones to extremes. Just use shades that pick up warmth from the solar energy you're getting free of charge.

To reflect heat and light down on the rooms' occupants, you can use a slick white ceiling. This will also provide a dramatic contrast to the deep tones used in the rooms' furnishings and walls. Backgrounds like this also lend themselves to the lavish use of exotic house plants, which will thrive in the warmth and sunshine. In gaily colored pots and tubs, they become stunning accents. And when you plan your indoor plantings, don't forget vegetables. Select the ones that are most pleasing to your eye and of a size suitable to in-house growing. Your sun-heated-and-lighted rooms make a perfect greenhouse that can put fresh garden produce on your table all winter. Herbs and flowers can be interspersed with them for a lovely effect, not to mention the economy. Simple and attractive planters to suit the décor of the rooms are easy to make, or may be purchased ready-made in a wide variety of styles.

Accessories in polished brass, such as lamps and wall plaques, show off well in such a setting, as do strategically lighted displays of crystal, coins, or trophies. Here your personal taste and style can assert themselves.

A word about drapes and curtains . . . Rods are available that extend beyond the side edges of the windows. These allow space against the adjoining wall area for the drapes when they are open. Thus, the entire window area is open to the sun during daylight hours. This window treatment also makes the windows seem larger than they are, a decorating ploy used by professionals even when solar heat is not a factor.

In summer, when the overhang shields your rooms from the sun's heat, the dark colors that were so serviceable and cozy in winter may not be suited to the general vacation mood. A change of décor is in order . . . without too much expense or labor. Slipcovers of cool-to-the-touch, easily cleaned or laundered materials in light, summery colors will make a stunning contrast against the dark walls and floors. And the floors do not have to retain their winter colors either. A few throw rugs in colors that echo the slipcovers' shades will transform the overall appearance of the rooms. If you like to capture the mood of the seasons, this is an easy way to have two completely different décors for very little outlay.

In all rooms with a southern exposure, where winter sun warms you during the daylight hours, heat can escape during the nighttime hours through the same windows that admitted it, because even insulating glass has less insulating power than the house wall. But you can block the heat escape with panels of expanded polystyrene foam. The panels are readily available in 2 foot $\times$ 4 foot size with an R factor of 4 in the usual 1-inch thickness. A 2-inch thickness with double the R value (insulating power) is also available, usually on order. The panels are easily cut to fit window openings, or joined and trimmed to fit larger window areas. The joining can be done with a special mastic made for foam. Do *not* use other types of mastic, as they may break down the foam. The foam can be framed with wood (as shown in Chapter 6) and painted with most water-base latex paints, or covered with wallpaper, using a water-base wallpaper paste. For a more formal effect, the panels can be covered with fabric. The lightweight wood framing and panel areas can be done in a Japanese motif to give them the appearance of shoji screens. These can slide away in shoji fashion, or fold out of the way like internal shutters, as your décor dictates. Your backup (fuel-burning) heating system will, of course, be needed at night, but it will operate for a shorter overall time because of the added insulating effect of

Applying expanded polystyrene foam panels to windows for increased nighttime insulation for air conditioning in summer, heating in winter. Panels are 1-inch thick, 2-feet wide and 4-feet long, have R value of 4. Two-inch thickness is also available. Cut panels to snug fit with hot stainless steel knife (heated by propane torch). Hot knife cuts it like butter. Panels may be friction-fitted or held with turnbuttons or other means. Panels can be put in place, removed in slightly more than 1 minute total. Drapes can be drawn over panels to conceal them completely.

the panels. If limited wall space prevents the use of folding or sliding panels, consider making a cocktail or coffee table that will look impressively thick and handsome, but will in fact be a panel container on legs; or make a combination chest and window seat to accommodate the panels when they are not in use. Either one is a double expense-saver for the smart homeowner. The top can be hinged for quick and easy access to the storage compartment. The storage units may be sized for the standard 2 foot × 4 foot panels or for larger joined panels so long as the overall dimensions don't exceed practical furniture size. For those who do not choose to decorate them, the panels can simply be cut to fit into the windows after sunset, and regular drapes can be drawn across the window expanse for a more conventional appearance.

An added bonus that comes from the south-wall sun-heated house is the overhang that shades it from the heat of summer. It is an ideal place for hanging planters, a delight to the eye anytime. Just be sure they are hung so that they cannot swing in close enough to hit and possibly crack a window. This is also a marvelous sheltered spot for bird feeders all year round. The overhang will keep snow and freezing rain off the feeder and its little patrons at their buffet.

When your house is so oriented on your property that the rooms facing south are not the largest or most used in the daylight hours, as living, dining, or playrooms, it is still possible to make better use of the light. Where direct sunlight is not available, as with a northern exposure, a fairly solid fence, painted a brilliant white, and set back far enough from the north wall of the house so that the fence catches the sun, can reflect a great deal of light into those windows that face north. Not all houses and properties are designed so that this can be done, but a few experiments, observing shadow patterns and propping up temporary fence panels, will soon tell you if this is practical. The sunlight not only adds cheer to north rooms, but reduces the need for electric lighting in the daytime, with resultant economy.

Since many solar homes are supplied year round with domestic hot water heated by the flat-plate collector on the roof, particular attention must be paid to the choice of trees around the building. Deciduous trees that shade the roof in summer should be placed in

Use of evergreens to shelter house from prevailing winds without blocking sunlight from south windows.

your landscaping plan so that they never shade that part of the collector that works all year, which would defeat the domestic hot-water segment of your system. With extra showers and laundry generated by summer activities, you'll definitely need the hot water. Considering the insulation installed in the roof of a solar house, a shade tree would do more good away from the house, where it will show off to a better advantage and provide a pleasant sitting area outdoors. On the northern side of a solar home, a row of evergreens, placed close enough so that they form a barrier against winter winds, actually acts as a set of "spoilers" in the wind. Aerodynamically speaking, they spoil the airflow of a strong wind, breaking it up into eddies and vortices that have much less wind-chilling effect. Carefully placed evergreens can also be used as winter windbreaks on the south side of the house (if that's where the prevailing wind comes

South windows shielded from sun in summer by vines and deciduous trees.

from) so long as they don't block light from the windows. As shown in the photo, they should not be much higher than the eaves of the roof, especially if roof collectors are to be installed. In addition to decreasing the wind-chill effect, evergreens serve as a very effective noise and dust barrier throughout the year. Use low-growing types under off-ground windows, and plan for pruning if necessary to keep your plantings from encroaching on other window areas. You want all south windows clear to admit full sun in winter. The overhang that should be above larger window areas will block the sun in summer. Drapes or ready-made awning-type overhangs (typically of aluminum) can be used for smaller windows.

Away from the house, your landscaping can follow a variety of patterns in areas not affecting your sun heating. An open expanse of lawn or garden may fit your plan. If you add a swimming pool, in

Vine that provides summer sun shield over this window is Virginia Creeper. It grows fast, sheds its leaves in winter, and withstands extensive pruning. Apple tree also shields these south windows.

ground or above, and plan to use flat-plate collectors in your sun-heating system later on, consider locating the pool fairly close to the house, where the collectors can be used to warm the pool water in spring and fall to extend the season. If privacy is a problem (as builders often place homes close to the road to shorten the drive-way), a locust pole fence or shield of evergreens placed far enough from the house so as not to block the sun from windows may solve the problem. Always remember, in planning, that the winter sun is lower in the sky and casts a much longer shadow.

Sun heat can also be put to work to aid in some aspects of your gardening. If your land is deep enough from north to south, an east-to-west wall of brick or stone, or of cement block tinted to a darker tone with cement coloring or paint to retain the heat of the sun, provides a sheltering backdrop that absorbs heat and so prolongs the gardening season for a surprisingly long time. The darker, more massive the wall, the more efficient it will be in storing heat. The wall should be built at the extreme northernmost edge of the property, so there will be sufficient open area to allow sun to reach wall and garden. Cold frames can also be built against the wall.

Flowering shrubs and ornamentals can, of course, be planted at will to landscape your grounds. These are simply some guidelines for making the most of the sun and its warmth.

As to heat collectors, if the south slope of the roof is away from the road, they will not show. If they do show, they may well become the latest status symbol and conversation piece, as did the TV antenna. Heat collectors can also be built into a fence that runs from east to west, the heat being carried into the house storage tank through insulated piping. This has already been done successfully in experimental installations. The fence, however, should not be far from the house. Such a fence might be part of a terrace enclosure.

Variations on the Trombe-Michel idea, described in Chapter 3, can be used for outbuildings such as unattached garages, boathouses (no mildew or rot problems), or other outbuildings where above-freezing temperatures are desirable. It's a great way to keep a seldom-used guesthouse from getting that dank, damp, mildewy atmosphere at no cost.

Index

Illustration Credits

Libbey-Owens-Ford Company: pages 3, 5, 10, 11; P.P.G. Industries (University of Delaware photo): 8; University of Wisconsin, Madison (from Engineering Experiment Station Report No. 21): 19, 20; UOP, Inc. (Wolverine Division): 33; National Woodwork Manufacturers Association, 44; Edmund Scientific Co.: 52; Solar Energy and Energy Conversion Laboratory, University of Florida: 58, 61, 62; Corning Glass Works: 62, 63, 162; Zomeworks: 67, 68; Hitachi Chemical Company America, Ltd.: 70; National Woodwork Manufacturers Association: 76; Owens-Corning Fiberglas: 80; National Cellulose Manufacturers Association: 82, 83; Dean Products Inc.: 98, 99; Revere Copper and Brass, Incorporated: 100, 101, 120, 125, 126, 130; Copper Development Association, Inc.: 109, 114, 143, 144, 161; Honeywell–National Science Foundation: 154, 156, 158. All other illustrations by George Daniels.